ROUTLEDGE LIBRARY EDITIONS: THE VICTORIAN WORLD

Volume 17

HEWETT COTTRELL WATSON

HEWETT COTTRELL WATSON
Victorian plant ecologist and evolutionist

FRANK N. EGERTON

LONDON AND NEW YORK

First published in 2003 by Ashgate Publishing Limited

This edition first published in 2016
by Routledge
2 Park Square, Milton Park, Abingdon, Oxon OX14 4RN

and by Routledge
711 Third Avenue, New York, NY 10017

Routledge is an imprint of the Taylor & Francis Group, an informa business

© 2003 Frank N. Egerton

All rights reserved. No part of this book may be reprinted or reproduced or utilised in any form or by any electronic, mechanical, or other means, now known or hereafter invented, including photocopying and recording, or in any information storage or retrieval system, without permission in writing from the publishers.

Trademark notice: Product or corporate names may be trademarks or registered trademarks, and are used only for identification and explanation without intent to infringe.

British Library Cataloguing in Publication Data
A catalogue record for this book is available from the British Library

ISBN: 978-1-138-66565-1 (Set)
ISBN: 978-1-315-61965-1 (Set) (ebk)
ISBN: 978-1-138-64337-6 (Volume 17) (hbk)
ISBN: 978-1-138-64342-0 (Volume 17) (pbk)
ISBN: 978-1-315-62936-0 (Volume 17) (ebk)

Publisher's Note
The publisher has gone to great lengths to ensure the quality of this reprint but points out that some imperfections in the original copies may be apparent.

Disclaimer
The publisher has made every effort to trace copyright holders and would welcome correspondence from those they have been unable to trace.

Hewett Cottrell Watson

Victorian plant ecologist and evolutionist

FRANK N. EGERTON

Foreword by
David L. Hull

ASHGATE

© Frank N. Egerton, 2003

All rights reserved. No part of this publication may be reproduced, stored in a retrieval system, or transmitted in any form or by any means, electronic, mechanical, photocopying, recording or otherwise without the prior permission of the publisher.

The author has asserted his moral right under the Copyright, Designs and Patents Act, 1988, to be identified as the author of this work.

Published by
Ashgate Publishing Limited
Gower House
Croft Road
Aldershot
Hants GU11 3HR
England

Ashgate Publishing Company
Suite 420
101 Cherry Street
Burlington
VT 05401-4405
USA

Ashgate website: http://www.ashgate.com

British Library Cataloguing in Publication Data
Egerton, Frank N.
 Hewett Cottrell Watson : Victorian plant ecologist and evolutionist.-(Science, technology and culture, 1700-1945)
 1.Watson, Hewett Cottrell 2.Plant ecologists-Great Britain-Biography
 3. Evolution (Biology)
 I.Title
 581.7'092

Library of Congress Cataloging-in-Publication Data
Egerton, Frank N.
 Hewett Cottrell Watson : Victorian plant ecologist and evolutionist / Frank N. Egerton.
   p.cm.--(Science, technology, and culture, 1700-1945)
 ISBN 0-7546-0862-X
   1. Watson, Hewett Cottrell, 1804-1881. 2. Plant ecologists--Great Britain--Biography. 3. Plant ecology--Great Britain. 4. Plants--Evolution--Great Britain. I.Title. II. Series.

QK31.W327 E43 2002
581.7'092--dc21
[B]
                                                                                    2002022628
ISBN 0 7546 0862 X

Printed and bound in Great Britain by MPG Books Ltd., Bodmin, Cornwall.

Dedicated to
David Elliston Allen
Andrea Lynn Egerton
Robert Glen

H.C. Watson (1846), by Margaret Sarah Carpenter.

Contents

List of Figures and Tables	xv
Foreword by David L. Hull	xvii
Preface	xxv
Acknowledgments	xxvii
Introduction	1
Psychobiography or Not?	2
Chapter Organization	3
PART I: FINDING A PLACE IN THE WORLD	5
1. 'As The Twig Is Bent, So Grows the Tree', 1804–28	7
2. Edinburgh and Career Possibilities, 1828–32	15
The City and Phrenology	16
The University and Plant Geography	24
PART II: THE LIFE OF A GENTLEMAN SCIENTIST	39
3. Relationships and Social Perspectives, 1833–59	41
Private Life	41
Family Life	49
4. Phrenological Struggles, 1833–40	55
5. Outlook and Social Responsibilities, 1835–60	67
Science and Religion	67
Political and Social Views	70

6.	Continuing Plant Geography Studies, 1833–48	83
	In Britain	83
	In the Azores Islands	86
7.	Relationship with William Hooker, 1834–50	99
8.	Seeking Employment, 1842–48	111
9.	Professional Relationships with Forbes, Babington and Balfour, 1833–59	121
	Conflict with Edward Forbes	121
	Botanical Colleagues: Watson versus Babington and Balfour	128
	Botanical Societies: Edinburgh versus London	133
10.	History Not Quite Repeated: Watson, the Botanical Society of London and *The Phytologist*, 1840–58	137
11.	The Origin and Transmutation of Species, 1832–47	147
	Watson's Orientation and Studies	147
	Darwin's Orientation and Studies	161
12.	Darwinian Parallels and Contrasts, 1809–58	163
	Early Life and Personality	163
	Early Thoughts on the Brain and the Emotions	164
	Early Studies on Biogeography	165
	Private and Family Life	166
	Outlook and Social Responsibilities	166
	Darwin and the Azores	167
	Involvements with Science and Scientists	168
13.	Stonecutter for Darwin's Edifice, 1847–59	177
	A Colleague for Darwin	178
	Watson's Own Scientific Conclusions	183
	Darwin's Use of Watson's Findings	190

PART III: LATER LIFE, WORK, AND INFLUENCES   195

14. Later Life, Work and Influences, 1860-81   197
    Doubts on the Darwinian Revolution   197
    A Synthesis on the Botany of the Azores   204
    Relationship with Joseph Hooker   208
    A Protégé for Watson   214
    Watson's Other Relationships and Influences   218
    Darwinian Parallels and Contrasts   230

Conclusions   233
    Personality, Education and Experience   233
    Scientific Achievements   238

Bibliography   241
    Abbreviations   241
    List of Publications by H.C. Watson   241
    Manuscripts   250
    Published Works   252

Index   281

List of Figures and Tables

Figures

Frontispiece	H.C. Watson (1846), by Margaret Sarah Carpenter. Royal Botanic Gardens, Kew.	
1	Congleton, drawn by Charles Wilson. Head (1887), frontispiece.	10
2	George Combe (1836), by Daniel Macnee. Gibbon (1878), I, frontispiece.	18
3	Hewett Cottrell Watson (1839), said to be drawn by Haghe, but bearing the initials 'R.D.D.' *The Naturalist* 4 (1839), facing p. 265.	45
4	George Combe (1857), by John Watson Gordon. Gibbon (1878), II, frontispiece.	78
5	Locator map of the Azores Islands. Godman (1870), facing p. 1.	87
6	Diagram of Pico, by C.F. Hochstetter. Seubert & Hochstetter (1843).	90
7	*Campanula vidalii* H.C. Watson. W. Hooker (1844).	96
8	William Jackson Hooker (1841), by Daniel Macnee. Royal Botanic Gardens, Kew.	102
9	Joseph Dalton Hooker (1855), by George Richmond.	106
10	Drawing by Ella Taylor: Princess Mary giving instructions to Craig, a gardener, Royal Botanic Gardens, Kew, Oct. 29, 1858, with Sir William Hooker and Ella Taylor. This is the image of Hooker that Watson held in later years. The Royal Collection © 2002, Her Majesty Queen Elizabeth II.	108
11	Edward Forbes holding a plant (1849), by T.H. Maguire.	124
12	Charles Darwin (1849), by T.H. Maguire.	170
13	Provinces, sub-provinces, counties and vice-counties of Britain. Watson (1847–59), III, frontispiece.	186
14	H.C. Watson (1871). Photo by Maull and Fox.	202
15	Map of the Azores Islands. Godman (1870), frontispiece.	205
16	J.D. Hooker (*c.*1870). Photo by Wallich.	210
17	John Gilbert Baker at work.	217

Tables

2.1	Watson's regions of British vegetation	35
2.2	Watson's delineation of upper and lower limits for two species at four locations	36
11.1	Watson's comparison of claimed species from six authorities	155

Foreword

The millions of species that populate the earth are related to each other in very complicated ways. They form all sorts of clusters, some sharply bounded, most with fuzzy, overlapping boundaries. Years ago I conducted a series of interviews with systematists and evolutionary biologists in which I asked my subjects which group of organisms they first investigated in any detail. I then asked if this particular point of entry into the living world influenced their later ideas. If they had gained entry to the living world by studying a different group of organisms, might they have come to different conclusions? All of my subjects said yes. Not only did their point of entry influence their early views, but also it probably continued to exert influence up to the present.

Victorian scientists also form a web of overlapping scientific relationships. One wonders what a study of Victorian scientists using the methods of numerical taxonomists would look like. Some scientists would be the focus of numerous professional alliances; other would be outliers. Unfortunately, students of Victorian science have not provided us with such a highly general mapping of Victorian scientists. Instead, we are presented with much more limited pictures. Some authors study groups of related scientists, for example, Rudwick (1985),[1] Secord (1986),[2] Richards (1987),[3] Desmond (1989)[4] and Oldroyd (1990),[5] but the usual mode of analysis remains the more tractable scientific biography. The reader gains entry into Victorian science through the careers of individual scientists. But if the point of entry into the living world has such a strong and lasting influence on systematists and evolutionary biologists, it is no less likely that the scientists who serve as points of entry for historians will influence our understanding of Victorian science.

---

[1] Martin J.S. Rudwick. 1985. *The Great Devonian Controversy: The Shaping of Scientific Knowledge among Gentlemanly Specialists*. Chicago: University of Chicago Press, xxxiii + 494 pp.

[2] James A. Secord. 1986. *Controversy in Victorian Geology: The Cambridge-Silurian Dispute*. Princeton: Princeton University Press, xvii + 363 pp.

[3] Robert J. Richards. 1987. *Darwin and the Emergence of Evolutionary Theories of Mind and Behavior*. Chicago: University of Chicago Press, xvii + 700 pp.

[4] Adrian Desmond. 1989. *The Politics of Evolution: Morphology, Medicine, and Reform in Radical London*. Chicago: University of Chicago Press, x + 503 pp.

[5] David R. Oldroyd. 1990. *The Highland Controversy: Constructing Geological Knowledge through Fieldwork in Nineteenth-Century Britain*. Chicago: University of Chicago Press, ix + 438 pp.

We come to know most Victorian scientists through their own life and letters as well as one or two later biographies, but the primary point of entry into Victorian science for most readers has been, and continues to be, Charles Robert Darwin. Literally hundreds of works devoted to Darwin and the Darwinian revolution have been published through the years. Might this imbalance influence how we understand Victorian science? Might not our understanding of Victorian science be enhanced if we cast our net more broadly? Desmond (1989) has contributed to this process by moving down from the dozen or so big guns at the time to lesser lights working on morphology and supported largely by their medical practice. Secord (2001),[6] shows us how Victorian science looked from the perspective of Robert Chambers. Now Frank Egerton in his *Hewett Cottrell Watson: Victorian plant ecologist and evolutionist* has done the same for Watson.

For anyone who has read much in the Darwin literature, the name H. C. Watson might sound somewhat familiar. Wasn't he a phrenologist? But as Egerton shows he was much more than a phrenologist. In fact, his work in botanical systematics and plant geography were of even greater importance than his early work promoting phrenology. When Watson began his career, Victorian science was becoming professionalized and a few scientists were even writing general treatises on the nature of science. Watson was not wealthy, but he possessed enough financial resources that he did not have to earn his keep. Although jobs in science that paid a salary were beginning to open up, Watson was never able to get any of these positions.

With the advent of inexpensive publication costs, professional journals were proliferating. Watson edited *The Phrenological Journal* for three years but became disillusioned by the poor quality of the papers that he received. They were not 'scientific' enough, at least not 'scientific' in the sense that philosophers of science such as John Herschel were urging upon scientists. But it should be noted that these same early philosophers of science rejected Darwin's theory of evolution and for the same reasons. Perhaps Darwin had fulfilled the requirements of discovery but not proof. All he did, as Herschel put it, was present his readers with the law of higgledy-piggledy.

Present-day readers are likely to be biased by the respective fates of phrenology and evolutionary biology. Both began as genuine scientific research programs, but phrenology degenerated into quackery while evolutionary biology succeeded in retaining its status as genuine science, periodic onslaughts notwithstanding. Watson fought against the descent of phrenology, but he lost. He thought that a science of

---

[6] James A. Secord. 2001. *Victorian Sensation: The Extraordinary Publication, Reception, and Secret Authorship of Vestiges of the Natural History of Creation.* Chicago: University of Chicago Press, xvi + 581 pp.

phrenology was possible. The problem was that no one, Watson included, seemed to have what it took to make the improvements in phrenology necessary for it to remain a genuine science. Watson repeatedly complained about this lack of ability in others. He was somewhat gentler with himself.

The part of phrenology that intrigued the general public most was the purported ability of phrenologists to read one's character from the lumps and depressions of one's skull. It was not the only 'science' at the time claiming the ability to make such inferences. For example, Johann Kaspar Lavater set out his science of physiognomy in the eighteenth century according to which the close examination of a person's face would allow inferences about this person's character. When Captain Fitz-Roy first met Darwin, he was put off by the shape of Darwin's nose and almost rejected him as a traveling companion. How different the history of science would be if Fitz-Roy's belief in physiognomy had been stronger. Upon Darwin's return from his voyage, his father exclaimed, 'Why, the shape of his head is quite altered.' Had Darwin's experiences during his voyage actually changed the shape of his head, and if so, did these changes influence his character? Darwin also remarked in his autobiography that a German phrenologist had concluded on the basis of a photograph of Darwin that he had the 'bump of reverence developed enough for ten priests.' Perhaps Darwin missed his calling after all.

Watson was a major figure in phrenology, but he had an even greater influence in botany. In attempting to make phrenology a genuine science, Watson was certainly struggling against the tide. Phrenology had too many trappings of a pseudo-science. But Watson found the task of improving the lot of botany, especially plant geography, equally daunting. These enterprises were in no sense pseudo-scientific. The worst that could be said is that they were pre-scientific. Many, probably most of the people who were engaged in botany were amateur collectors, keeping in the mind that the distinction between amateur and professional was only emerging at the time. Watson spent a lot of his own time and money adding to his collections and working with other collectors, but he envisioned more. He searched for patterns in the distribution of species in the British Isles, an endeavor that today we would call biogeography.

Here the data overwhelmed Watson. One might say that if he, like A.R. Wallace, had come to accept evolution before he began his investigations, he might have made the same sort of contributions in biogeography that Wallace did decades later. This assumption, though plausible, has one drawback. Watson had accepted the transformation of species from at least 1834. Darwin took his voyage seeing every phenomenon from the perspective of independent origins. Watson and later Wallace had the advantage of making their observations from an evolutionary perspective. Wallace succeeded in discovering major patterns in biogeographic distributions; Watson was much less successful.

When Darwin decided to be less secretive about his own views on evolution, he invited four of his fellow scientists to Down House on 22 April 1856. The four authorities that Darwin chose were not only reasonably open-minded but also could help him to improve his theory – Joseph Hooker, T.H. Huxley, Thomas Vernon Wollaston and H.C. Watson. His choice of Hooker and Huxley was obvious. He had worked in close collaboration with them for some time. The choice of Wollaston was less obvious, but he had just published a long monograph on the variation of species. In a review of the *Vestiges of the Natural History of Creation* in 1845, Watson declared his position on the evolution of species. Hooker brought this review to Darwin's attention, noting that Watson was 'almost an avowed believer in Progressive development' (CCD 3:211). In his reply to Hooker, Darwin does not mention Watson. How much Watson's belief in the transmutation of species had on Darwin's decision to invite him to the meeting in Downe is hard to say. Watson's wide knowledge of plant geography might well have been sufficient in itself.

Hooker, Huxley and Wollaston accepted Darwin's invitation; Watson did not. And here we confront a major theme in Egerton's biography – the influence that Watson's acerbic personality and reclusive habits had on his career in science. Egerton is well aware that psychohistory is currently out of fashion, but if any character in science cries out for some attention to the influence that his personality had on his role in science, Watson's does. He hated his father. Quite a few sons do, but not only was his hatred extreme, he was willing to acknowledge it after his father's death. No disingenuous platitudes for Watson. One frequently hears that hypocrisy is the lubricant of society. No one can say that Watson was a hypocrite. He was blunt to a fault. He made very few professional friends, and those he did make usually did not last long. To stay on good terms with Watson took an incredible amount of maneuvering. William Hooker was one of the few who succeeded.

How much did Watson's irascible demeanor have on the reception of his views? After reading Egerton's biography of Watson, I am forced to conclude that it made a lot of difference. If you are super bright as Newton was, and royalty to boot, scientists are willing to make allowances. If they can make use of your work in their own research, they are willing to put up with quite a bit of eccentric behavior, but what if others are producing work similar to yours? What if your work is only good enough? Then other scientists need not put up with the emotional wear and tear that results from interacting with you. What if Watson had attended Darwin's 1856 meeting? What if he was a good old boy who could laugh it up at meetings of the X Club? Then I suspect that today we would think very differently of him and his contributions to science.

No historian enjoys spending his life studying a person whom he does not like. Once I asked Bill Coleman[7] why he ceased working on Cuvier when his 1964 book appeared. He answered that he could not stand the man. We owe a great debt to Frank Egerton for spending so many years studying a man who was anything but lovable. We can all benefit from his perseverance.

David L. Hull
Northwestern University

---

[7] William Coleman. 1964. *Georges Cuvier Zoologist: A Study in the History of Evolution Theory*. Cambridge, Massachusetts: Harvard University Press, x + 212 pp.

Preface

Some years ago, I traveled to Cambridge University to study the vast collection of Darwin's manuscripts and correspondence located in the Library's Manuscript Department.

Rather quickly I stumbled upon a batch of letters from a botanist, Hewett Cottrell Watson, of whose existence I was unaware. These letters were written during the few years preceding publication of Darwin's book, and they very perceptively answered questions relating to evolution. I was quite surprised, because there was a substantial body of literature on the history of evolutionary biology. How could such an important correspondent have been overlooked?

I was an American historian of biology primarily interested in the history of ecology and evolutionary biology. Had I been British, I might have known that the scientific journal of the Botanical Society of the British Isles is named Watsonia, in Watson's honor. He was actually well known to botanists, but had not been the subject of a book-length biography. As I extended my studies on him, I became intrigued both with his ecological studies and his personality. My book on Darwin receded into the background as I indulged my curiosity about a classic Victorian eccentric. I was astonished at how much I could learn about this relatively neglected figure.

Acknowledgments

During more than a quarter century of working on this book, I have become endebted to many people and institutions for assistance, and I extend my thanks to all.

David Elliston Allen, of the Wellcome Institute for the History of Medicine, has been of enormous assistance. His interest in Watson long pre-dates my own, and his extensive knowledge and advice have been invaluable all along the way. He willingly read and critiqued the whole manuscript – some chapters several times – and his own published scholarship has been among the most important guides to the development of my own ideas. To whatever extent the text does not seem too Americanized for British tastes, the reader can thank him for stylistic intervention. He also assisted me in my search for a publisher.

My wife, Andrea, has provided moral support – and in some summers financial support – and her advice has been important for handling all sorts of non-scholarly aspects of the project.

Professor Robert Glen, Department of History, University of New Haven, has also read and critiqued the whole manuscript. I am fortunate in having a friend who has a professional interest in British history that includes Watson's lifetime. Indeed, Glen had even done research on Holland Watson, but his assistance extends far beyond such a happy coincidence to include all sorts of advice on my handling of Watson within his society.

Professor David L. Hull, Department of Philosophy, Northwestern University, has also provided valuable comments on the whole manuscript, from both historical and philosophical perspectives. The Victorian era has provided a happy hunting ground for case histories in many of his philosophical studies, and I am fortunate to have his dual expertise on which to draw. He also assisted me in my search for a publisher.

Three other scholars have read those portions of my manuscript relating to their expertise. Doctor Jean-Marc Drouin, Muséum National d'Histoire Naturelle, provided valuable advice on Watson's contributions to phytogeography, ecology, and his use of works by French-language botanists. Professor Michael Mitchell, Department of Botany, University College, Galway, not only advised me on Watson and Irish botany, but he and his wife, Brigid, served as our tour guides around Ireland. He also advanced my studies with his gifts of four books, including one by Watson. Professor Michael Gurtman, Department of Psychology, University of Wisconsin-Parkside, read my discussions of Watson's personality and saved me from making pronouncements beyond the limits of my expertise. My book owes much to their comments.

Financial support for my researches and writing enabled me to complete this book. The University of Wisconsin-Parkside gave the most support, in the form of a sabbatical leave, summer salary, and research funds. The University of Wisconsin Graduate School also funded one summer's research. Both the National Science Foundation and the American Philosophical Society provided funds for travel to Great Britain. My first summer of research in Cambridge, England, was funded by the Hunt Institute for Botanical Documentation.

Libraries have been both major resource centers for my work and also places where I received expert assistance from librarians and archivists. It would be impractical for me to attempt to itemize all these debts, but I feel compelled to acknowledge here the three British institutions where I worked most extensively; the others are listed under 'Manuscripts' in the bibliography. Peter J. Gautrey, who was head of the Darwin Archive, Cambridge University Library, during the summers I worked there, both welcomed and enlightened all who came to use that wonderful collection. The Cambridge University Libraries were a great place to launch my project, because wherever my curiosity took me, I invariably found what I needed in their stacks. At the Royal Botanical Gardens, Kew, I primarily used the Archives, but its Library was also available to me for checking on published details that came up while I read in the archives. The same was true at the National Library of Scotland, repository of the George Combe Archive. In America, many libraries have sent me books and photocopies of journal articles on interlibrary loan. On my side of the Atlantic, I single out for special thanks – because of heavy use – the University of Wisconsin-Parkside Library and its staff and also the collections at the University of Wisconsin-Madison.

Introduction

During the Victorian era, science changed from a largely part-time amateur activity during the 1820s into a highly professional enterprise by the 1870s. The Darwinian Revolution in biology, in which Watson participated, was a major component of the increasing scientific activity. He was also a pioneer in a new science not yet even defined – ecology.

Watson was practically the first naturalist to conduct researches on plant evolution, beginning in 1834. (Erasmus Darwin had made a few observations on it, but even if his poetic speculations are considered science, they had no apparent influence on his contemporaries.) Ironically, Charles Darwin made better use of Watson's data than Watson himself did. Watson was one of the four leading experts on species whom Darwin invited to Down House in 1856 to discuss the species question. There is further irony in the fact that Darwin's generous acknowledgment to Watson in *The Origin of Species* (1859) obscured rather than revealed his debt. It was only when Darwin's long manuscript on natural selection was published in 1975 and his correspondence with Watson was published, beginning in 1988, that Watson's contributions became clear.

In plant geography and ecology, Watson's contributions were also forgotten, not through anyone's fault, but because they were overshadowed by Darwin's domination of British biology and Watson's lack of any distinctive theoretical discovery. He, like Darwin, followed in Alexander Humboldt's footsteps, though he went to more trouble than Darwin did to measure environmental factors in the hope of relating them to species distributions. Watson also hoped that his data could have a bearing on the evolution of species. In that respect, his wish came true, though it required his playing Brahe to Darwin's Kepler.

Although various botanical projects occupied most of Watson's scientific endeavors, at the beginning he was torn between botany and phrenology, and for a dozen years it seemed that phrenology might win. He found it attractive because it provided an alternative to his father's ideology, but he also wanted to understand his own psyche. He was a more daring theoretician in phrenology than he ever was in botany, but in 1840 he gave up phrenology because he could not raise it to the level of accepted science. Nevertheless, he retained phrenology as a component of his personal ideology.

Watson not only conducted scientific investigations, he also attempted to advance both phrenology and botany within their social organizations. His efforts on behalf of phrenology failed because phrenology itself failed. However, botany was already an established science, therefore he was more successful in helping organize British botany. Because The Botanical Society of the British Isles continues to advance his goals, it calls its journal Watsonia.

Psychobiography or Not?

Watson's activities in science were accompanied by conflicts with several colleagues. Yet if he had had a more tranquil personality, he might have followed his father into a legal profession and not worked in science at all. When writing the biography of someone who suffered from psychological problems, the biographer might present the protagonist as eccentric and merely provide a narrative account without interpretation, or he might become an amateur psychiatrist and provide interpretations based upon limited evidence. Biographies of Isaac Newton provide examples of both approaches.

Frank E. Manuel first confronted Newton in his *Isaac Newton, Historian* (1963), which shows Newton as a superb rationalist who nevertheless worked under some non-rational preconceptions. Manuel's psychobiography, *A Portrait of Isaac Newton* (1968), interprets, among other things, Newton's frequent conflictual relationships with rivals. Finally, Manuel published a series of lectures on *The Religion of Isaac Newton* (1974). None of these books escaped criticism by skeptical reviewers, but taken together, they nevertheless represent the most substantial description of a scientist's personality and intellect. Yet Manuel did not delve into the sciences on which Newton's reputation rests. More recently, historian of science Richard S. Westfall has written a detailed, authoritative biography of Newton which deals with his scientific career and achievements. Westfall's response to Manuel's *A Portrait of Isaac Newton* is that it 'is subtle, complex and ingenious', but 'is it also true? It appears to me that we lack entirely any means of knowing.'[1] Newton had more conflicts with other scientists than Watson did, but when A. Rupert Hall wrote a history of the most famous conflict – with Leibniz over who invented calculus – he focussed on the intellectual and social aspects of the situation, without delving into Newton's psyche.[2]

Rather than choose between the approaches of Manuel on the one hand or of Westfall and Hall on the other, I am writing a biography that deals with both the personality and the scientific contributions. What Watson achieved in science is more important than understanding his personality. Yet Watson's personality

---

[1] Westfall 1980, 53, n. 36.
[2] Hall 1980. Both Hall and Westfall have continued to publish on Newton, but without changing their basic perspectives; Dobbs 1994.

influenced his career, and failing to deal with this matter neglects an important part of the story. In doing so, I am encouraged by the example of Stephen Gaukroger, whose intellectual biography of Descartes (1995) treats personality as an essential ingredient of scientific achievement, and by Jane Camerini, whose review of two biographies encourages this kind of a biography of scientists.[3]

Watson left behind an ample record of his personality problem. Both his childhood situation and his adult behavior seemed to me compatible with the diagnosis of a specific disorder. However, Professor Michael Gurtman responded to some of my relevant chapters with the comment that my biography does not provide a rigorous enough context or analysis to establish a psychodiagnosis. Nor does he think my account would be enhanced significantly even with a valid diagnosis. What is important is to show how Watson's personality problems both oriented him toward science and limited his effectiveness as a scientist.

Chapter Organization

There are three chronological parts which correspond to important changes in Watson's life. Part I covers Watson's childhood, youth, early maturity and education. There is enough biographical information for us to gain insights into both his personality development and his achievements. Chapter 1 surveys his childhood experiences, which oriented him away from his father's outlook and profession, and made him receptive instead to phrenology and botany. Chapter 2 follows him to Edinburgh, where he expected to make phrenology, and maybe also botany, his profession.

Part II encompasses the diverse aspects of his life and work after he left Edinburgh in 1833 until he and Darwin published their seminal treatises in 1859. The fact that both books appeared in that year was coincidental, but their books were related in the sense that Darwin was picking Watson's brain while they were writing on similar subjects. Part II is the heart of this biography, because it deals with Watson establishing himself as an independent scientist and with the main contributions which he made to plant ecology and evolution. Along the way, he interacted with his peers and his family in both positive and negative ways. These interactions are explored to clarify why he was able to achieve as much as he did, and also why he did not achieve more.

Two older men with whom he had become acquainted in Edinburgh, George Combe and William Hooker, served as his mentors during many of these years.

---

[3] Camerini 1997, reviewing biographies of T.H. Huxley by Adrian Desmond and of Richard Owen by Nicolaas A Rupke. For additional discussions of biographies of scientists, see Shortland & Yeo 1996.

However, the gratitude he felt did not save them from his criticism, which he viewed as a purely intellectual matter, not to be taken personally. That was the same attitude, only more so, that he displayed toward colleagues whom he viewed as rivals, including Edward Forbes, Charles C. Babington and John H. Balfour. That Darwin was exempted from his biting criticism is interesting evidence of Watson's more generous impulses. Chapter 13 explores the striking similarities between Watson and Darwin's lives and work.

Part III concerns the later years of Watson and Darwin. Both men continued their scientific research, writing, and interactions with colleagues as long as their health lasted. Darwin's achievements and fame far outshone Watson's, yet the parallels and contrasts between them remain interesting even in this period.

PART I
FINDING A PLACE IN THE WORLD

Chapter 1

'As the Twig Is Bent, So Grows the Tree', 1804–28

> I never knew an individual towards whom I felt such a permanent and bitter antipathy as to my own father.
>
> Watson to G. Combe, 9 July 1848

Personality is often a crucial factor in the degree of one's success or failure in life, and so it was for Hewett Cottrell Watson. He was born into a home which anyone would have judged to be very favorable. His father was a member of the upper middle class who took very seriously the discharge of his responsibilities – as he saw them – to family and society. Both Holland and Harriett Powell Watson were upstanding, religious citizens. Their child-rearing practises were probably typical of their class, and probably similar to those they had experienced as children themselves. Although they had ten children, this large family need not have seemed a liability, because they had servants.

From birth, Hewett was specially favored, as the first son in a family that practiced primogeniture. He alone would inherit enough wealth to lead an independent life. He should therefore have been the happiest of the children, but instead he was the unhappiest. Hewett came to hate his father and to feel out of harmony with his siblings. What went wrong?

Holland Watson (d. 10 March 1829, aged 79) was an attorney who practiced in his native Stockport by 1781 and served there as Justice of the Peace in the 1790s. The events of his life seem reasonably consistent with Hewett's later caustic characterization of him: 'A Tory magistrate, full of the prejudices of his party; most violent in temper; opposed to all philanthropical measures; and almost dead to individual benevolence.' Holland Watson was a zealous administrator who wanted to stifle any dissent before it could reach major proportions. Thus, when English Jacobins began to win a small following among Stockport workers in 1792, he led the way in establishing a local branch of the ultra-conservative Reeves Association, which was dedicated to suppressing radical movements. His command of the Stockport Volunteer Corps must have provided a tangible means for him to disperse those gatherings which he deemed seditious. Yet he preferred to attack opponents in the local press. He maintained the status quo, but perhaps he expected the opposition to collapse more readily than it did, since he later exercised more caution in his attacks than he had at first.[1]

Since Hewett hated his father, his testimony alone would be insufficient to establish Holland Watson's qualities. The latter revealingly summarized part of his own career in 1822 while recommending his brother-in-law, John Lloyd, for a government position to the younger Robert Peel. With 'active zeal' Lloyd had helped Magistrate Watson disperse 'seditious and tumultuous meetings', and bring 'to justice the seditious and disaffected persons in the populous towns and neighborhood of Stockport'. The high point of these actions came on 3 May 1801, when Watson, Lloyd and Rev. William Robert Hay dispersed 'a seditious meeting of about 10 000 persons, held at Buckton Castle ...'. Therefore, Watson recommended Lloyd 'not only as being eminently qualified for the situation by his professional knowledge; but as having (in my judgement) strong claims on government for the sacrifices he has made ...'.[2]

The Watsons' seven daughters were all born before their three sons. Hewett was born on 9 May 1804, after the family moved to Park Hill, near Firbeck, in Yorkshire. Shortly thereafter they moved again, five miles or so, to Worksop in Nottinghamshire. When Hewett was three or four, they finally settled in or near Congleton. His earliest memory was of this move: he worried about how to get Toby, his father's favorite horse, to their new home. Hewett's solution was 'the brilliant conception that Toby should ride on the top of the carriage'.[3] Was Hewett actually worried that he, not Toby, was in danger of being left behind? It is unlikely that his father would get angry at a favorite horse, but if Hewett was not as 'good' as Toby was, he might have worried about whether his father wanted to take him to the new home.

The younger brothers were born after the family reached Congleton. Holland, the father's namesake, was born in 1809, while his father was mayor, and Zeph in 1811.[4] Hewett's special status was eroded, and it is unlikely that the Watson household ran as smoothly after their births as it had before. His brothers competed for their mother's love and attention. Still Hewett's main source of love and security, she became an

---

[1] Hewett Watson to George Combe, 22 November 1834, and even stronger, 9 July 1848. Glen 1984, chaps 3 and 6. Holland Watson's will mentions his role in the militia.
[2] Holland Watson to Peel, 28 July 1822; British Library, Add. Ms. 40,348, f. 191. Robert Glen supplied Hay's first two names and the date of the incident at Buckton Castle (which is a hill seven miles northeast of Stockport, not a castle).
[3] Watson 1839c, 266. This sketch of his early life was ostensibly written by the editor of *The Naturalist*, but is transparently autobiographical.
[4] Holland Watson was mayor of Congleton in 1809–10, and Deputy Mayor in 1820; Stephens 1970, 341. See also Head 1887.

imperfect haven. She undoubtedly protected her younger sons from his hostile acts, and she may have informed her husband when Hewett needed disciplining.

Hewett's bitter comments about his father indicate his belief that as a child he was subjected to a tyranny which left permanent psychic scars. Twenty years after his father's death, the intensity of his hatred was undiminished:

> I never knew an individual towards whom I felt such a permanent and bitter antipathy as to my own father. We were totally unsuited to each other. My mother, a widely different character, died when I was fifteen, just when the child's fear of his father was changing into dislike and even hatred. I do not recollect that we ever had an actual quarrel after that age; but I shunned all sort of intercourse or communication which could be avoided. Ever since, I have felt that my own mind, and course of after life, were very injuriously affected by him. So much so, that were I to indite an autobiography, the injury done to the son by a father intensely irascible, thoroughly unreflective, large-brained, feeble in benevolence and conscientiousness, strong in veneration, directed only to political and social (not religious) matters, vain personally just of those qualities which the son was deficient of, etc., etc., the injury done to the son by the father would be exposed in the most unreserved, and probably even vindictive manner ...[5]

Hewett's youthful sibling jealousy apparently changed over the years to indifference or contempt. Many years later, he could still remember the emotional hardships which he experienced as a teenager without his mother, but he never noticed that it was even harder for his brothers to grow up without their mother, for they were even younger than he when she died. His sisters should have pointed this out to him, but probably there was too little closeness between them for this to happen. His sisters were not necessarily afraid of him, but if he had some authority over their younger brothers, they might have deferred to him in that regard.

Hewett's contempt for his brothers was probably reinforced by their adoption of their father's values. Any inclination which they might have felt toward rebellion would have been blocked by Hewett's hostility. He, not their father, was their enemy, therefore it would have been Hewett's values which they rejected. If they knew of it, it would have seemed unfair to them that the brother who detested their father was nevertheless destined by tradition to inherit most of the family wealth. Probably their main consolation in their last years at home was knowing that, when of age,

---

[5] Watson to Combe, 9 July 1848. Harriett Watson died 16 March 1819, aged 50, two months before Hewett was 15. She is buried in the graveyard of St Peter's Church, Congleton; inside the church Holland Watson placed a memorial tablet 'to perpetuate the memory of the best of wives'.

Figure 1 Congleton, drawn by Charles Wilson.

they could escape into the army. Their father arranged to purchase commissions for both of them in the East India Company's regiments.⁶

One possible escape from an unhappy family life is school. For Hewett, however, the Congleton Grammar School represented stultifying drill in the Bible and in Latin grammar, and he acquired 'a most intense hatred' for both. After a few years he left – with the reputation of being 'an incorrigible dunce'. He fared better in the Rev. Isaac Bell's school in Alderley, where he stayed six years. With none of the beatings he had endured previously, he rose to the top of the class. Although Hewett still found the subjects taught (mostly Greek and Latin) pedantic and irrelevant, Bell lent him books on science and history to read after school.⁷ His father had books on horticulture, agriculture, natural history, medicine, and some volumes of the *Philosophical Transactions of the Royal Society of London*, but if Hewett read any of them, he did not mention it later.⁸

The sole companion from his youth whom Hewett thought worth mentioning in his autobiographical sketch was the family gardener. This relationship:

probably contributed, in connection with his mother's taste for floriculture, to give young Watson a partiality for flowers; as a child at home, and subsequently as a schoolboy, his garden was the chief amusement of his playhours.

Gardener and garden saved him, he thought, from 'the family taste for Topography, Heraldry, Antiquities, and other things of the past'.⁹ Watson later recalled that his 'boyish fancy for plants and floriculture ... attracted the favorable notice of Dr. Stanley, whose opportune instruction and encouragement gave a scientific direction to the taste, and rendered it the solace and relief of the child during a period of protracted bodily suffering'.¹⁰

The 'protracted bodily suffering' to which Watson alluded was due to an accident. At age 15 he was hit on his right knee by a cricket bat during a game. Since the 1950s it has been possible to replace a crushed knee joint with an artificial one, but in his day no such help was available.¹¹ Painful though the accident was, it saved

---

⁶ The father's will (28 March 1828) indicates these arrangements. Information on their service and deaths is at the India Office Library and Archives, London.
⁷ Watson 1839c, 266–7. Isaac Bell's daughters were remembered in a will which Watson wrote about April 1842.
⁸ After Holland Watson's death, an auctioneer compiled a catalog of saleable possessions, which listed books in his library. Branch & Son 1829.
⁹ Watson 1839c, 266.
¹⁰ Baker 1881; quoted from Watson 1883, x. When Watson knew him, Stanley was Rector of Alderley Parish.
¹¹ Watson to G. Combe, 22 November 1834. Sonstegard et al. 1978.

him from the military career his father had planned for him, for which he was temperamentally ill suited. His disability never prevented him from hiking through the mountains of Scotland or ascending Mount Pico in the Azores, and his belief that its impact was more emotional than physical may have been correct.

The Rev. Dr Edward Stanley was an unusual parish priest, who later, in 1837, became both bishop of Norwich and President of the Linnean Society of London (and retained both offices until his death). He may have served as an alternative father figure for Hewett. The contrast was between an authoritarian magistrate who broke strikes at the mills and a liberal churchman who helped his social inferiors to better themselves.[12] Stanley had a serious interest in the sciences of ornithology, entomology, mineralogy and geology as well as in educational, political and religious reforms. Stanley also raised three sons, two of whom attended school with Watson. Later, Watson remembered that he had given cakes to Arthur Penrhyn Stanley and had 'interfered for his protection'.[13] He had no such recollection concerning his own brothers.

Since Hewett was repelled at the thought of having to learn any more Latin or Greek at Oxford, a Church career was unrealistic. His father accordingly apprenticed him, at age 17, to Messrs Jackson, a firm of solicitors in Manchester. Watson became friends with the Quaker furrier and silk hat manufacturer Joseph Eveleigh, who was also the city's leading field botanist. Manchester in the 1820s was fast becoming a major industrial, economic and scientific center, an exciting place for bright young men.[14]

Nevertheless, after two years Watson 'tired of the town or of the Law, and transferred himself to Liverpool, where he remained about eighteen months'.[15] Watson could be very precise when he wished, and his comment that he abandoned a career and a city because one or the other displeased him is strangely ambiguous. We should not assume, however, that he left Messrs Jackson in disgrace, since if he had, he probably would not have mentioned the law firm by name in an autobiographical sketch published in 1839.

His motivation for leaving Manchester was not the only ambiguity in his sketch. Presumably he was 19 when he left there, which would have been in 1823; if he then

---

[12] On Stanley (1779–1849), see Prothero 1898b.

[13] Baker 1883, x. A.P. Stanley became Dean of Westminster and a widely respected man of letters. Owen Stanley became a naval captain of the surveying ship HMS *Rattlesnake*. Desmond 1997, see index; O'Byrne 1849, 1109; Prothero 1898a and c.

[14] Allen & Lousley 1979, 163, n. 9; Inkster & Morrell 1983; Kargon 1977, chap. 1; Thackray 1974. Watson mentioned Eveleigh (1786–1838) in his first letter to William Hooker, 9 November 1830. For a brief sketch and references on Victorian Manchester, see Glen 1988.

[15] Watson 1839c, 267. On the legal profession during his apprenticeship, see Orth 1988.

remained in Liverpool only eighteen months, he would have left it about 1825. But the next place he tells us he lived was Edinburgh. Since he went there to enter its university, which he did in the autumn of 1828, it is unlikely that he arrived three years earlier. Most likely he went back to Congleton, but to say so in print might have seemed to be an admission of failure.

However, Congleton could have had no attraction for him, and in the autumn of 1825 the family moved to Liverpool. It is surprising that a feeble Holland Watson would have undertaken this move so late in life. Perhaps he did so to broaden the cultural life of his children still at home. However, the strain was greater than he had anticipated, and he wrote to his friend, the Rev. Joseph Hunter, on 23 January 1826, that he might not live to see another January: 'You can have no idea of the change I have undergone within the last twelve months. I am sunk into an old man all at once, altho' I want some years of Eighty. My mental as well as corporeal faculties are greatly impaired ...'.

In this weakened condition, he may have turned to his oldest son and main heir to help manage the family until the younger sons were old enough to join the army. Holland Watson probably saw nothing permanently significant in his earlier conflicts with his oldest son, seeing them as merely part of growing up. He probably had conflicts with his own father, which as an adult he forgot. Surely it did not occur to him that his main heir could hate the hand that would feed him throughout a lifetime. As Holland Watson struggled to finish raising his many children, he may have thought that Hewett had become less quarrelsome, and therefore more mature. Aside from this, Holland Watson's final tasks were to find husbands for his daughters, buy commissions for his two younger sons, and to draw up his will.[16]

In spite of an aversion to his father, Hewett was probably active in the family's move to Liverpool. It was the sort of task – clear-cut and limited in duration – in which he could succeed at age 21, and that success would have reassured his father in believing that he could also rely upon him for help in raising his two youngest sons. In that task Hewett could not succeed, because it required sympathy and patience in addition to firmness, and he could only provide the last of these qualities when relating to his brothers.

Liverpool was more congenial for Hewett than anywhere else he had lived until then. The city had grown wealthy largely from the slave trade, but its prosperity continued after abolition. Its 'clean' commerce made it more stylish than 'dirty' Manchester, whose prosperity came from manufacturing. Hewett would have heard of, and perhaps met, Liverpool's most revered citizen, William Roscoe

---

[16] Holland Watson to Joseph Hunter, 23 January 1826; British Library, Add. Ms. 24,876, f. 290–1. His will is dated 28 March 1828, a year before he died. He is buried in the yard of St Michael's Church, Liverpool; Head 1887, 196.

(1753–1831). Roscoe was an attorney and banker who had more than a local reputation as a historian, poet and botanist. He was also a strong opponent of slavery, and helped abolish it in England during his one term as a Whig MP in 1807.[17] He and Watson shared an interest not only in gardens and botany, but also in liberal reforms.

In Liverpool, Watson acquired an interest in phrenology. Hostility toward an orthodox father made him receptive to unorthodox ideas. Dr George Douglas Cameron was most responsible for arousing this interest. Cameron, while a student at the University of Edinburgh Medical School, had been a 'tallish, rather slender, handsomely formed, fair, very sinewy fellow; apt at all exercises; a fair scholar, not extensively read, but singularly prompt with all he did know; cheery, witty, affable, and consequently popular'. He had become acquainted with phrenology before receiving his degree in 1822; he then settled in Liverpool. In 1824 he offered to lecture on the subject at Liverpool's Royal Institution. His offer was declined, because 'phrenology is not yet sufficiently established as a science'. (Two years later the Royal Institution did allow the new Phrenological Society to use one of its rooms for meetings.)[18]

Despite some failures, Hewett still had confidence in his intellectual abilities, and his pending inheritance would provide the resources which he needed to prepare for a profession of his own choosing. With all his sisters married off, or soon to be, and with Zeph finally packed off to the army, he could resume his education, this time in Edinburgh.[19]

Watson leaves the impression that after eighteen months in Liverpool he was excited enough about phrenology to depart for Edinburgh's medical school himself. This impression is strengthened by a later memoir of him: 'When he was about twenty-two, the bequest of a small estate in Derbyshire from a member of his mother's family placed him in what for a man of his simple habits were independent circumstances. Upon this he removed to Edinburgh ...'[20]

---

[17] For a biography and color portrait of Roscoe, see Chandler 1953. For his botanical activities, see Desmond 1977, 530. For early natural history activities, see Greenwood 1980, 376–7, and Kitteringham 1982.

[18] This would not be the last time the phrenologists were denied a meeting opportunity. Cameron practiced in Liverpool long enough to be listed in the city directory for 1829, but not long enough to earn mention in Bickerton's *Medical History of Liverpool*. Bickerton 1936; Christison 1885–86, I, 135–6. See also Stephens & Roderick 1972, 65.

[19] In a letter to G. Combe, 22 November 1834, he wrote that he 'left my father's house & went to Edinburgh nominally as a student of medicine, really to see and know Phrenologists'. This seems to indicate that he left before his father died.

[20] Watson 1839c, 267; Baker 1883, x–xi.

Chapter 2

Edinburgh and Career Possibilities, 1828–32

... the discovery of Phrenology [can] be the greatest blessing ever placed within the grasp of man ...
>Watson to W. Alison, as quoted to G. Combe, 1 March 1830

While the distribution of plants, in most of the mountainous countries of Europe, has engaged the ... attention of philosophic naturalist ... in our own country this department of Botany has been almost utterly neglected ...
>Watson 1832, iv

Watson rejected his father's career but wanted a career of his own choosing. He also rejected his father's Toryism and Anglicanism, but found a substitute for both in phrenology. It was his main reason to seek a university education, though phrenology was an uncertain career possibility. There were no professorships of phrenology, and the best tactic seemed to be to combine phrenology and medicine. This is what the founders of phrenology had done, and Dr Cameron was one of the few physicians who had followed in their footsteps. Watson intended to do the same, but this decision did not exclude the simultaneous pursuit of his botanical interests, because medical botany was part of the curriculum.

The very act of striking out on his own in a new city with a vibrant intellectual life, centered around its university, was exciting. Edinburgh was Britain's foremost scientific and medical center. The city had three scientific societies, and the university had a natural history and a medical society. Watson arrived in the autumn of 1828 and found lodgings in the heart of the city at 27 Howe Street, just off Royal Circus, midway between the university and the Royal Botanic Garden. Looking downhill from his address, one sees the Firth of Forth in the distance, while close by are several attractive parks. Today it is a stylish neighborhood of terraced houses and cobblestone pavement.

The City and Phrenology

Phrenology is remembered as a pseudoscience. Its leaders strove long and hard to achieve scientific credibility, and partial success kept their endeavor alive throughout the 1800s. It was far from being a minor fringe movement, for it attracted such notable adherents as John Quincy Adams, Prince Albert, Alexander Bain, Honoré de Balzac, Paul Broca, Charlotte Brontë, Robert Chambers, Henry Clay, Auguste Comte, George Eliot, David Ferrier, Robert FitzRoy, Karl Marx, Clemens Metternich, Edgar Allen Poe, Étienne Geoffroy Saint-Hilaire, Herbert Spencer, Mark Train and Alfred Russel Wallace.

Its founder, Franz Joseph Gall, while a medical student, thought he had discovered a correlation between excellent memory and protruding eyes. He received an MD degree at Vienna in 1785 and then used his broad clinical experience in the hospitals, prisons and insane asylums to list supposed correlations between skull shape and mental attributes. Previously, the mind had been studied by philosophical speculation, and the brain had been studied by dissection. Science had not made a detailed connection between mind and brain, and Gall took up this challenge.[1]

He demonstrated some localized functions in the brain in lectures, 1796–1801, and he might have achieved modest recognition as a neural physiologist had that been his goal. His attempts to do far more than that brought both disciples and controversy. His most active disciple was Johann C. Spurzheim, whom Gall won over in 1800.[2] The Austrian government feared the materialist implications of their doctrines, and finally forbade further teaching of phrenology. They went on a two-year lecture tour through Germany, Switzerland, Denmark and Holland, arriving in Paris in 1807. In Paris, they lectured and dissected brains before several learned societies.[3] Hoping to be elected members of the Institut de France, they submitted to it a memoir on their anatomical researches on the brain. A committee appointed to evaluate it found some things to commend, but cautioned that 'no conclusion can be drawn from [their memoir] concerning a doctrine which has only a very distant relation to anatomy'.[4]

---

[1] Lanteri-Laura 1970; Lesky 1970; G. Richards 1992, 170, 257–64 *et passim*; Young 1972. The list of notable phrenology adherents comes from Hull 1978, 137.

[2] Walsh 1975.

[3] Ackerknecht 1958; Lanteri-Laura 1970; Temkin 1947 and 1953. Some discussions from Gall and Spurzheim, *Anatomie et physiologie du système nerveux* are published in English in Clarke & O'Malley 1968, 393–4, 476–80, 599–02, 826–7.

[4] Tenon, Sabatier, Pinel & Cuvier 1809; see 37. The memoir was mainly by Cuvier and originally appeared in *Mémoires de la classe des sciences mathématiques et physiques de l'Institut de France*, (1808) 109–60.

In antiquity, Galen had originated the experimental technique of vivisectioning the brain and spinal cord of dogs and pigs, but later investigators seldom continued the practise until the 1800s. Luigi Galvani began electrical experiments on nerves in 1791, and by the early 1800s Pierre Flourens and François Magendie were conducting vivisectional studies on the nervous system in Paris, as was Sir Charles Bell in England.[5] Gall knew that comparative anatomy, physiology and pathology could clarify the functioning of the brain, and some such studies were included in *Anatomie et physiologie du systême nerveux en général, et du cerveau en particular*, which he began publishing in 1810, assisted at first by Spurzheim. However, their detailed anatomical studies ended about that time. George Henry Lewes discussed the history of phrenology in 1867, and observed that not a single phrenologist since Gall and Spurzheim had contributed to anatomy or even kept abreast of its progress.[6] Phrenological methodology depended entirely upon observations and correlations. Although these methods can be useful, scientific advancement is slow when it neglects experimentation.

Since phrenologists shunned experimentation, they needed some other reliable method. One possibility was medical statistics, which Pierre Charles Alexandre Louis and his disciples began developing in Paris during the 1820s.[7] However, phrenologists relied on only a few case histories when drawing correlative conclusions, and never collected enough data to have statistical significance. Their few measurements of skulls were merely to determine average sizes, and they had no way to verify whether the averages really indicated what they claimed.

Gall argued for the existence of 27 'organs' on the cerebrum and other parts of the brain, which controlled as many faculties. Spurzheim obtained his MD degree at Vienna in 1813, and then spent the remainder of his life lecturing, mostly in Britain. His treatise, *The Physiognomical System of Drs. Gall and Spurzheim* (1815, 571 pages + 18 plates), raised the number of mental faculties from 27 to 33, and left open the possibility that others might be discovered. When he died during an American tour in 1832, he had already passed the torch to the Combe brothers.

George Combe (1788–1858) was one of 17 children of an Edinburgh brewer. Poor health and a dreary grammar school did not bother him as much as the idea of Jesus having died for his sins; he 'envied the cattle that had no souls'. Becoming interested in the mind as a teenager, he read works by leading British philosophers, but without seeing how to apply their principles to life. He developed a busy law practise, but found time to take a course in human anatomy and physiology. He was disappointed that Dr John Barclay made no insightful connections between these

---

[5] Clark & O'Malley 1968, 178–83, 291–303.
[6] Lewes 1867, II, 411. Lanteri-Lauri 1970, 75.
[7] Delaunay 1953; Underwood 1951.

Figure 2 George Combe (1836), by Daniel Macnee.

subjects and mental processes.[8] Still hopeful, he attended Spurzheim's dissection of a brain in 1816. He studied Spurzheim's doctrines, and was converted. He collected skulls and casts and soon began lecturing ardently on phrenology. In 1819 he published the first edition of his *Essays on Phrenology*, which he steadily enlarged and revised, retitling the fifth edition *A System of Phrenology* (1843). Along the way, he gradually increased the number of human mental faculties from 33 to 45. In spreading the phrenological gospel, Combe was assisted by his several relatives, most notably by his brother Andrew (1797–1847), who passed surgery exams in 1817 and received an MD degree from Edinburgh in 1825. Andrew wrote on phrenology from 1823 to 1846, and also gained a reputation as a medical author with his *Principles of Physiology Applied to the Preservation of Health* (1834), which went through 15 editions.[9]

The Combe brothers and four others founded the Edinburgh Phrenological Society in 1820. A year later it had 33 local and corresponding members. When its membership reached 80, George Combe persuaded the Society that prospective members should henceforth submit to phrenological examination to determine suitability for admission. Despite this precaution, he continued to be disappointed in the quality of the membership.

In 1823, Andrew Combe addressed the Royal Medical Society of Edinburgh on 'Does Phrenology Afford a Satisfactory Explanation of the Moral and Intellectual Faculties of Man?' The ensuing debate lasted for hours, with both the phrenologists and the Medical Society claiming victory. The Combes had the entire debate transcribed in shorthand, and requested permission to publish it, but the Medical Society would not agree. The Phrenological Society then began publishing *The Phrenological Journal*, and Andrew's address appeared in its first volume.[10]

This episode was one of many occasions when phrenologists apparently were not bested in debate but were nevertheless silenced by established authority. Another example occurred after George Combe debated Sir William Hamilton in 1827–28; Hamilton refused to publish his lectures.[11]

---

[8] Gibbon 1878, I, 39, 92–3; de Giustino 1975; Richards 1992, 264–5, 326–8 *et passim*; Walsh 1971, 269–78.

[9] George Combe 1850, 50–1, 136. Andrew's contributions to P*hrenological J.* are listed 553–7 and to *The British and Foreign Medical Review*, 560.

[10] Gibbon 1878, I, 129, 141, 144, 161. A volume of transactions appeared in 1822, but not as a regular periodical. The history of this society is summarized in Watson 1836a, 129–33.

[11] Combe 1850, 130–3; Gibbon 1878, I, 191–203. Cantor 1975a and 1975b, and Shapin 1975 debated the extent to which these Scottish struggles turned on social versus intellectual issues; both aspects were important. On Hamilton, see Athey 1988, Cooter 1984, 285, and Stephen 1890.

External hostility sometimes strengthens a group, when its members have strong common interests. A test for the Edinburgh Phrenological Society came when George Combe explored the relationship between phrenology and his liberal outlook, and expressed his conclusions in a tract which he distributed to members of the Society for comment. 'Bewilderment, horror, and indignation took possession of many of his best friends', and he was accused of infidelity. William Scott published an 81-page rebuttal, maintaining that Combe's views on natural law were 'fundamentally erroneous and contrary to sound natural reason and to sound Scriptural doctrine'.[12] Combe made revisions, and in 1828 published his most popular work, *The Constitution of Man Considered in Relation to External Objects*. He explained human behavior as caused by physical qualities and education, and never referred to God. Although he maintained that he was religious, the devout of Edinburgh were unconvinced; Scott and many others fell away. The Society seemed on the verge of collapse.[13]

At this point Watson entered the picture. He studied at the Edinburgh medical school from autumn 1828 to December 1831. The Combes were pleased to have a bright young medical student join their ranks. Watson contributed to phrenology as both scientific investigator and advocate, but he left the university without a degree. Consequently, for him phrenology could not be an aspect of a medical career, as it was for Andrew Combe and a few other physicians.

Only the first of Watson's studies was clinical. He wrote it after one year in medical school, and the girl he studied, like one reported by his mentor, Dr Cameron, was allegedly deficient in the faculty of Tune.[14] Watson's report now seems embarrassingly superficial, yet it was acceptable to *The Phrenological Journal*:

> I had occasion to visit the house of a farmer, and, immediately on entering it, was struck with the extreme deficiency of Tune in the head of his daughter, a child aged four years ... her mother informed us that she was quite unable to speak, and asked if we thought the hollow at the region of Tune had any thing to do with her inability ... Not being competent to give the numerical development adopted in Edinburgh, I can only state in general, that the head was small, and tolerably well-balanced in the occipital, basilar, and sincipital regions; Cautiousness and Love of Approbation very strong, Self-esteem greatly deficient, Imitation fair, the anterior lobe small in proportion to the rest of the head by reason of its narrowness, though pretty full; Individuality and Eventuality large; Language moderate: Tune, besides

---

[12] Gibbon 1878, I, 181, 186. Combe's tract was On *Human Responsibility as Affected by Phrenology* (1826).
[13] Gibbon 1878, I, 181, 236. Cooter 1976 and 1984, part 3. Grant 1965. Parssinen 1974.
[14] Cameron 1825. On Cameron, see Cooter 1984, 277.

the narrowness of the forehead, greatly depressed. She was very curious, and evidently much pleased with a watch and other objects shown to her, handling and examining them with great attention. She appeared to hear and understand, nearly as well as children of that age usually do, what was said to her, and her mother told us, that she had never observed any defect in these particulars. She once acquired the power of pronouncing 'dad,' but lost it again.

For medical readers it may be proper to remark, that the temperament inclined, though not very strongly, to lymphatic, and the child appeared delicate. The mother said her bowels were irregular, and within the last year she had occasionally had fits.

He had hoped to examine her again the next summer, but she died of smallpox a few months after his visit.[15] In this report he attempted to extend the data of phrenology rather than to test the validity of phrenological theory.

Students, Thomas Kuhn tells us, learn science from textbooks that serve as paradigms or guides to the methods and standards of the discipline. A paradigm should represent a substantial achievement, and its appearance indicates the maturity of a science. It directs attention to the kinds of problems that seem worth investigating, and explains means of solving them.[16] Ostensibly, Gall had published the paradigm for phrenology. In 1836 Watson claimed that phrenology was a better constructed science than botany.[17] However, its structure was built with anecdotal data that could not withstand critical examination. As Lewes commented later, phrenology had reached too hasty a 'maturity'.[18] Watson was familiar with texts by Spurzheim and by Combe, and he often cited Gall as well, without specific reference, because his knowledge of Gall's writings came from the extensive quotations in English contained in the first five volumes of *The Phrenological Journal* (1824–28).[19]

After his first school year at the University of Edinburgh he returned to Liverpool and drafted a constitution for a new phrenological society. He sent a copy to George Combe (28 June 1829), explaining that: 'The two last regulations are to give the spirit of reality and liberality to the society and to prevent religious disputation and biblical evidence.' He credited Spurzheim, who attended the meeting, with preventing the rejection of Watson's regulations.[20]

---

[15] Watson 1829. He mentioned the girl's death in *Statistics of Phrenology* (1836a), 196.
[16] Kuhn 1970, chaps. 2–5. Ambiguities in his paradigm concept are discussed by Masterman 1970. Kuhn's subsequent development of his concept is discussed by Hoyningen-Huene 1993.
[17] Watson 1836a, 9.
[18] Lewes 1867, II, 408–12.
[19] Watson mentioned to George Combe on 28 February 1838 that he had not, until very recently, seen Gall's work since he left Edinburgh.
[20] The Phrenological Society of Liverpool was established in May 1828. Watson 1836a, 146.

Watson's approach to organizing the Liverpool Phrenological Society was influenced both by what he had observed in Edinburgh and by his father's managerial style. His father had used the public press to further his causes, and Watson appreciated the advantages of this strategy. He saw an urgent need to advance both phrenology and the public's appreciation of it, but he advised Combe (14 July 1829) that public education was the more pressing problem, and *The Phrenological Journal* should emphasize that.

Phrenology was excluded from university curricula, but that was not unique. Many scientific innovations, from Copernican astronomy to Darwinian evolution, had been similarly excluded for some years before winning acceptance.[21] Watson was therefore in a similar situation to the disciples of other controversial sciences when he commented to George Combe (25 February & 1 March 1830) that his physiology professor, William Pulteney Alison, had both inadequate knowledge of brain functions and an unfair bias against phrenological theory. (Alison was a leading physician and medical professor.) Watson did absorb enough of an orthodox perspective to disagree with fellow phrenologists about the importance of experimental vivisection.[22]

During that summer, he again became like a missionary from a mother church and wrote to Combe (30 August 1830) in Edinburgh about the prospects for a phrenological society in Manchester. There was hostility among leaders in science:

> amongst whom is John Dalton, the chemist, president of the L. & P. Society, who recently said from his chair that 'he did not know there was such a *Science* as Phrenology'.

Watson had called on about a dozen residents who felt favorably toward phrenology, and he expected that 'a Society in Manchester would soon flourish, if once well established'. His banker friend, Thomas S. Scholes (probably a stepson of Watson's sister, Harriett Anne Watson Scholes), was one of those contacted, and Watson later dedicated his *Statistics of Phrenology* to him in appreciation for his support.[23]

Watson's concern over the small number of phrenologists was shared by the Edinburgh leadership. Some still blamed Combe's *The Constitution of Man* for driving people away. He retorted that phrenological discussions must be relevant to people's lives. His arguments persuaded one patron, who made a bequest to

---

[21] Barber 1961. Kuhn 1970.
[22] Watson to George Combe, 21 January 1832.
[23] Watson's statement in *Statistics of Phrenology* (1836a), 153, that the Manchester Society had been established in 1829 is thus a misremembrance.

subsidize the printing of both *The Phrenological Journal* and 70 000 cheap copies of *The Constitution of Man*.[24]

George and Andrew Combe did teach phrenology classes outside of university, and phrenology could thus be mastered as rapidly as other sciences – or *more* rapidly, because the training was less rigorous. While still a neophyte in 1830, Watson published two theoretical articles. Their purpose, in Kuhn's terms, was partial paradigm reformulation.[25] Precedent came from the texts of Spurzheim and Combe, for Spurzheim had revised Gall's concepts and Combe had revised Spurzheim's. Watson's two articles – 'Evidence Towards Ascertaining the Real Function of Comparison' and 'Inquiry into the Function of Wit' – argued cases more elaborately than was usual in either of these treatises. Previously, William Scott had also published articles on both faculties in *The Phrenological Journal*.[26]

Watson's article on comparison drew correlations between qualities of literary works and the shapes of heads of authors, the latter being determined from busts, face casts or portraits. He chose his subjects according to the principle that 'The more unequal is the cerebral development of any individual, the stronger and more easy evidence does he afford to the phrenologist, in respect to particular powers.'[27] His argument was that the poet Thomas Moore had a large organ for Comparison, and Alexander Pope had a smaller one; these differences caused Moore to use more similes in his poetry than Pope; however, the differences could not be explained entirely as a simple ratio between the faculties of Comparison, because the tendency to use similes appeared to be influenced also by other faculties – Individuality, Tune, Eventuality, Form, and Size. Such discussions as this led Lewes to ask how phrenologists could be confident of their answers when they claimed an interaction of different 'organs' since physicists had not yet solved the much simpler three-body gravitational problem.[28] Phrenologists were aware of this challenge, and drew a parallel with the chemist's dealing with elements and compounds.[29] Watson acknowledged that he lacked enough evidence to delimit the faculty of Comparison,[30] and recognized the value of seeking counter-evidence.[31]

---

[24] G. Combe to P. Neill, 18 July 1831, quoted in Gibbon 1878, I, 240. Details of William R. Henderson's bequest are in Gibbon 1878, I, 255–63. The cheap copies were printed by Robert Chambers.
[25] Kuhn 1970, 19.
[26] Scott 1827a and 1827b.
[27] Watson 1830a, 384. See also Watson's letter to G. Combe, May 1830.
[28] Lewes 1867, II, 417.
[29] Anon. 1824b.
[30] Watson 1830a, 394.
[31] 'Somewhat more than mere facts and illustrations is necessary before a *principle* can be firmly established; we must not only seek facts in support of it, but whether there be others contradictory.' Watson 1830a, 383–4.

The University and Plant Geography

While Hewett Watson was in medical school at Edinburgh, students were still studying botany because plants were important sources of medicines, and the test for a medical license in England and Wales still included a test of botanical knowledge. He made important contacts at Edinburgh among botanists as well as among phrenologists. Although his interest in plants began in childhood, his interest in plant geography – a rather academic subject – developed while he was a university student.

Descriptions of both habitat and geographical range of plants are included in Theophrastos's *Historia Plantarum*, both for Greek plants and others sent back to Athens by scholars traveling with Alexander the Great during his conquests. However, early botany was much more descriptive than theoretical, and little progress was made toward a science of plant geography until the mid-1700s.[32] The Swedish naturalist Carl Linnaeus (1707–78) was interested in what we call ecological and biogeographical questions. Essays by Linnaeus and others in the later 1700s prepared the way for a science of plant geography.[33]

The main founder of plant geography was the German polymath, Alexander von Humboldt (1769–1859). Although a far greater scientist than Gall, he was less successful in writing a paradigm treatise for his new science than Gall was for his. Humboldt's limitation was partly the sparseness of phytogeographical data, but mostly the vastness of his scientific interests. If he had limited himself to plant geography, he might have produced a better paradigm treatise than he did. Neither he nor anyone else developed a very sophisticated methodology for this subject. He did develop a new perspective, emphasizing the study of vegetation (correlations between environmental factors and the distribution of plant forms) more than traditional floristics (which species live in particular areas). Humboldt and his traveling companion, the botanist Aimé Bonpland, spent five years exploring Latin America, then went to Paris and published an epochal series of scientific treatises.[34]

Their *Essai sur la géographie des plantes* (1807; German edn 1807) was a good step toward paradigm production. Plants should be considered in relation to: the different zones and heights they inhabit; the degrees of atmospheric pressure, temperature, humidity and electrical charge of the atmosphere in which they live,

---

[32] On the history of early botany, see Greene 1983, with a discussion of information from Alexander the Great's expedition on 203–4 & 469–70.

[33] Browne 1983, 16–42. Du Rietz 1957. Larson 1986. Linnaeus 1972.

[34] Biermann 1972. Browne 1983, 42–52 and 59–62. Cannon 1977, chap. 3. Castrillon 1992. Dettelbach 1996. Drouin 1995. Grisebach 1872b. Nicolson 1987 and 1990. Stafleu & Cowan 1976–92, I, 274–6, II, 363–71.

and whether they are solitary or social species. One should study assemblages of plants, such as those species which form extensive heaths. He explained how terrain and climate enable the same plants to occur continuously from Canada to the mountains of Mexico, and how the Mediterranean Sea and the Pyrenees Mountains prevent the spread of African and Spanish plants into the rest of Europe.[35]

Humboldt believed that when a number of species are found on adjacent shores of discontinuous land masses, those lands must have once been joined. He even hoped that plant geography would 'be able to show with some certitude the first physical state of the earth'.[36] Plant geographers should also investigate 'whether among the immense variety of plant forms one can recognize some primitive forms and whether the diversity of species ought to be considered the effect of an accidental degeneration of varieties which has become permanent with time'.[37] However, Humboldt did not think this possibility very important.[38]

The *Essai sur la géographie des plantes* has a large folding chart of two volcanic mountains in Ecuador – Chimborazo and Cotopaxi – the sides of which were divided into districts according to the ranges of species; the Latin names of species are listed on the chart at the elevations where they occur on the mountains. Each side of this chart has vertical columns of figures which indicate: elevation above sea level, refraction when seen from a distance, elevation in comparison with other mountains, electrical phenomena in relation to cloud level, crops grown, progressive decrease in gravitational attraction, intensity of the sky's blueness, progressive decrease in humidity and barometric pressure, maximum and minimum temperatures, chemical composition of the air, height of lowest limit of perpetual snow in different latitudes, kinds of animals found, degrees centigrade at which water boils, geological features and sunlight intensity. One can correlate the magnitude of any of these environmental factors with the plants listed on the sides of the mountains. This correlation presumably indicates the physical requirements for any species.[39] He hoped that laws would emerge after others constructed charts similar to his for different parts of the world.[40] His science of correlations is called 'Humboldtian science', and his methods were widely adopted for decades.[41]

Humboldt's methodological deficiency was similar to that of Gall and his disciples, who were trying to erect a science of phrenology upon a series of correlations. Yet, if one compares the two endeavors, one sees that Humboldt's had

---

[35] Humboldt & Bonpland 1807, 14–18.
[36] Humboldt & Bonpland 1807, 20. Von Hofsten 1916, 262–3.
[37] Humboldt & Bonpland 1807, 20.
[38] Humboldt & Bonpland 1807, 27.
[39] Humboldt & Bonpland 1807, 42–3.
[40] Humboldt & Bonpland 1807, 78.
[41] Brock 1993; Cannon 1977, chap. 3.

the virtue of carefully obtained data, whereas Gall's relied upon insufficient or uncritically collected data. Gall was better at designing a science, and Humboldt was more careful about building materials. Gall erected a precarious structure that came tumbling down in a storm of skepticism, and Humboldt erected a sturdy, partial structure which endured but was incomplete. In later works, Humboldt discussed other aspects of plant geography without solving his methodological problem.

Göran Wahlenberg (1780–1851) followed Linnaeus's example by becoming Professor of Botany at Uppsala University and by conducting botanical and geological exploration in Lapland. His phytogeographical researches were strongly influenced by Linnaeus's and Humboldt's writings.[42] Wahlenberg's three books are about the plants of Lapland, Switzerland and Hungary; each contains a long introduction on the plant geography of the region followed by a list of its species with their 'Hab.' (habitat or habitation). Each introduction divides the book's region into distinctive zones, for which he provided descriptions and lists of species. He carefully collected data on elevations and monthly temperatures, and indicated which species were unique to the region and which had broader ranges. He was masterful in constructing charts and maps which provided important information without the distraction of Humboldt's botanically insignificant data, such as an index of the sky's blueness at different elevations. Wahlenberg failed to establish the relative importance for plant distribution of various environmental factors, but in several ways his approach to plant geography seems more realistic than Humboldt's.[43] Watson's *Outlines of the Geographical Distribution of British Plants* (1832) and several later works were organized similarly to Wahlenberg's monographs.

In 1820 Augustin Pyramus de Candolle (1778–1841), who worked in Geneva and became one of the most prominent Continental botanists,[44] wrote an excellent survey article, 'Géographie botanique', for the *Dictionnaire des sciences naturelles*. Linnaeus influenced his organization of the science under three headings and various subheadings: Influence of the Elements (Heat, Light, Moisture, Soil, Atmosphere), Stations of Plants (Marsh, Meadow, Forest and so on) and Habitations (Europe, Siberia, India, and so on). At first sight, De Candolle's plant geography seems static, reflecting the pre-evolutionary mentality in botany. There were, however, two potentially dynamic aspects. First, he stressed historical changes relating to geology, and second, under the heading of Habitations he discussed the different means by which plants are disseminated from their place of origin. That

---

[42] Eriksson 1976. Stafleu & Cowan 1976–92, VII, 17–21.
[43] Wahlenberg 1812, 1813 and 1814.
[44] Browne 1983, 52–64 et passim. Drouin 1991, 208–12, 1994, 77–89, and 1995. Pilet 1971b. Stafleu & Cowan 1976–92, I, 438–52.

discussion included the first consideration of competition as an important biogeographical factor.[45]

Yet De Candolle's survey article was no more a complete paradigm for plant geography than Humboldt and Bonpland's *Essai* was. Humboldt stressed broad goals and techniques of measurement; de Candolle provided a superior overview. Taken together, they offered a good introduction and points of departure. It was perhaps too early in the history of this science for an adequate paradigm to emerge, defining worthwhile problems and indicating how to solve them.

In Great Britain, botany, like science as a whole, was almost entirely conducted by amateurs, and a nebulous line separated serious studies from the hobby of plant collecting. John Ray had published a list of the plants of the Cambridge area in 1660, and many other local lists had been published subsequently, but these were insufficient for defining the ranges of species.[46] In 1805, Dawson Turner, a wealthy Yarmouth banker, and Lewis Weston Dillwyn, a Swansea pottery manufacturer and civic leader, published *The Botanist's Guide through England and Wales* (2 vols). Both men were Fellows of the Linnean and Royal Societies, and their prominence in science and society enabled them to collect botanical data by means of a printed questionnaire. Their guide listed for each county all known species, where found, and by whom.[47] Such precise information had never been available before for such a large area. However, the arrangement of information was for collectors, not phytogeographers, and no phytogeographic conclusions could be drawn from their guide.

Robert Brown (1773–1858), whom Humboldt called *botanicorum facile princeps* ('unquestioned leading botanist'), studied medicine at the University of Edinburgh, served as naturalist on Flinders's Australasian Expedition, 1801–5, then spent the rest of his life in London as an officer of the Linnean Society and also of the British Museum. He wrote two studies which compared the floras of Australia and the Congo with other near and distant regions. His 'General Remarks, Geographical and Systematical, on the Botany of Terra Australis' (1814) was based upon his own collection of some 4000 species during the voyage of the Investigator, whereas his Congo report (1818) was based upon a less extensive collection of

---

[45] Candolle 1820 was translated into English by William J. Hooker and absorbed into a longer article (with acknowledgment), Hooker 1834. Candolle 1820 and Hooker 1834 are reprinted in Egerton 1977. On the importance of Candolle 1820 for competition theory, see Egerton 1971. Watson seems never to have appreciated this insight, either from reading Candolle or Lyell.

[46] On the history of amateurs v. professionals, see Allen 1985. On the history of local floras, see Gilmour 1963.

[47] On Turner (1775–1858) and Dillwyn (1778–1855), see Desmond 1977, 622 and 187; Stafleu & Cowan 1976–92, I, 657–9 & VI, 544–6.

plants made by Christen Smith before he died during his expedition.[48] Brown devoted most of these reports to the distribution of the families of plants, each family being discussed individually. He analyzed the percentages of Dicotyledon, Monocotyledon and Acotyledon species of these regions in comparison with the percentages of each of these divisions in Europe. Differences in the proportions which were found in Europe, Australia and the Congo he explained in part by differences in climate.[49] In comparing Congo plants with species known from the shores of America and India, Brown encountered an interesting puzzle: 16 or more Congo species seemed almost, but not quite, identical with those found elsewhere. No one could study plant geography for long without having to face the question of the nature of a species.

Amateur botanist Nathaniel John Winch (1768–1838) and zoologist William MacGillivray (1796–1852) continued and enlarged the tradition of compiling local Floras. Winch's *Essay on the Geographical Distribution of Plants, through the Counties of Northumberland, Cumberland, and Durham* (1819; edn 2, 1825) was devoid of theory, yet it may have influenced Watson as much as some of the more comprehensive phytogeographical works. Winch provided a miscellany of information: geographical features of these counties, the number of species according to families, comments on whether economically useful species were native and abundant, the kinds of crops that grow best in these counties, and comments on the plants under ten headings.[50] The three counties which Winch examined are mountainous and contain two-thirds of the species found in the British Isles. MacGillivray, a native of Aberdeen, was in Edinburgh while Watson was a student there, being in charge of Professor Robert Jameson's museum, 1820–31, and then Conservator of the Museum of the Edinburgh College of Surgeons, 1831–41. Watson might have been present at the meeting of the Wernerian Natural History Society on 2 April 1831 when the Secretary read MacGillivray's 'Remarks on the Phenogamic [*sic*] Vegetation of the River Dee, in Aberdeenshire'. His paper described the topography of that river valley and its plants at different elevations – useful information, but adding nothing to phytographic theory.[51]

---

[48] On Brown, see Desmond 1977, 96; Mabberley 1985; Stearn 1970; Stafleu & Cowan 1976–92, I, 364–70. On Christen Smith (1785–1816), see Desmond 1977, 565.

[49] Brown 1814, 537–40; reprinted in Brown 1866, I, 8–10. Brown 1818, 423; reprinted in Brown 1866, I, 103. See also Browne 1983, 62.

[50] Winch 1819 and 1825. Edn 2 is organized differently to Edn 1. On Winch, see R. Desmond 1977, 669; Stafleu & Cowan 1976–92, VII, 362–3. Watson's letters to Winch in 1832 show an appreciation of Winch's contributions to plant geography.

[51] MacGillivray 1832. On him, see Desmond 1977, 406; Stafleu & Cowan 1976–92, III, 221–2.

William Jackson Hooker became Professor of Botany at Glasgow University in 1820, and the leading authority on the plants of Scotland.[52] His *Flora Scotica* (1821) followed the example of Turner and Dillwyn's *Guide*, and gave for each species a description and range, with the range including exact locations and authorities, thereby achieving a high degree of geographical precision. Hooker's *British Flora* (1830) was a manual of broader scope intended for identification work by students.

Watson's fondness for plants since childhood led him to begin compiling a herbarium and making excursions to seashore and mountainside in quest of specimens even before he moved to Edinburgh. In England, he probably used Turner and Dillwyn's *Guide* when planning these trips, and after moving to Edinburgh Hooker's *Flora Scotica* must have become his botanical bible.

When Watson attended it, Edinburgh University had a natural history society for students, the Plinian Society, which Darwin joined just before Watson's time, but Watson did not join it.[53] He did join the University's Royal Medical Society. The city had a College of Surgeons, which maintained the museum that employed MacGillivray. Professor Robert Jameson had built up the university's natural history museum to outstanding proportions,[54] and he also edited the *Edinburgh Philosophical Journal*, which was devoted to all branches of science. Medicine attracted almost half the University's students, there being in the school year 1825–26 a total of 2013, of whom 902 were in medicine. Of these, 250 were English.[55]

Watson as a diligent student. George Combe later remembered that he was:

characterised by steady application, great aptitude in acquiring knowledge, and a comprehensive power of appreciating its relations. He devoted his attention to the various branches of science usually included in the curriculum for a medical degree, and was elected by his fellow-students President of the Royal Medical Society of Edinburgh, in which office he distinguished himself by superior abilities and attainments.[56]

Fellow student J. Hutton Balfour had similar memories.[57] Combe and Balfour were writing letters of recommendation for Watson, but their recollections are confirmed

---

[52] References on W. Hooker are in chap. 7, n. 1.
[53] I am indebted to Miss Lucy Robertson, Assistant, Deptartment of Manuscripts, Edinburgh University Library, for having checked the Plinian Society's minute book and the University matriculation record for Watson's name. On Darwin and the Plinian Society, see Ashworth 1935, 102–5; Darwin 1959a, 50–1; A. Desmond & Moore 1991, 31–4.
[54] Sweet 1970a and 1970b on the museum. The botanic garden is discussed below.
[55] Ashworth 1935, 98. Chitnis 1986. Grant 1884, I, 329. Rosner 1991.
[56] G. Combe, 12 August 1846; quoted in Watson 1846a, 26.
[57] Balfour, 19 January 1842, quoted in Watson 1842a, 9.

in general terms by Professor Robert Christison, who recalled that 'presidents of the Royal Medical Society have always been the *élite* of the advanced students and young graduates'.[58] Professor Graham in 1832 spoke of Watson as 'my excellent, intelligent, and quietly persevering friend'.[59] Except in botany, it is unclear what Watson absorbed of permanent value from his medical studies. Yet he left Edinburgh with fairly definite ideas about science, some of which must have come from his studies.

Watson's opinion of his professors, excepting Graham, is generally unknown. His low opinion of Alison was apparently because of his opposition to phrenology. Since Watson was difficult to please, we may assume that his evaluations would have been no more enthusiastic, and perhaps less objective, than those of three contemporaries who commented on some of the professors.[60] Alexander Monro III, who taught anatomy and surgery, was competent but dull: he read his grandfather's lecture notes. James Home had once been a popular professor of *materia medica* but had changed professorships at age 63 (in 1821) and was never competent in his new chair, clinical medicine. Darwin wrote that 'Dr. [Andrew] Duncan's lectures on Materia Medica at 8 o'clock on a winter's morning are something fearful to remember.' James Hamilton was energetic, knowledgeable, and the leading authority on obstetrics in Scotland; and though vituperative, he was popular with students. Equally popular was Thomas Charles Hope, whose chemistry lectures were as famously interesting as Monro's were dull. Robert Jameson's lectures were very dry, and he gave Darwin a temporary aversion to geology. Yet, his classes were popular, because most students respected his great energy and devotion to his work, and they enjoyed the museum and field trips.

Watson's most significant training at Edinburgh was under Professor Robert Graham (1786–1845). He had received his MD degree at Edinburgh and became the first Professor of Botany at the University of Glasgow. He made an excellent impression there, and he advanced to the Professorship of Botany and Medicine and Keeper of the Royal Botanic Garden at Edinburgh in 1820. Three years later he persuaded the university to purchase a larger site for the Garden at Inverleith Row.[61] Because in 1820 his salary was only £100, Graham was forced to practice medicine as well as teach and supervise the garden, and he found little time for research. He collected materials for a book on the flora of Scotland which he never wrote. His only publications were notes in the *Edinburgh New Philosophical Journal*,

---

[58] Christison 1885–86, I, 162.
[59] Graham 1832, 351.
[60] Christison 1885–86, I, 67–90. For Darwin, see refs. in note 53. Wilson & Geikie 1861, 88–121. Payne 1885 is a very favorable account of Alison. See also Morrell 1972.
[61] He spent much time and effort there and was assisted by a capable curator, William McNab. Desmond 1977, 262 on Graham, and 412 on McNab. Fletcher & Brown 1970, chap. 9 on Graham and 8 on McNab. Ransford 1846.

published four times a year from July 1824 until April 1837 on new or rare plants which were found around Edinburgh. He also published a few reports on plants collected on summer field trips. Although George Wilson remembered Graham's classes as among the most popular in the university, he did critize him for slowness in abandoning Linnaeus's artificial classification for de Jussieu's more recent natural system, and for not keeping up with progress in knowledge of plant anatomy, physiology, reproduction and distribution.[62]

In 1827, Graham began conducting annual week-long field trips in August or September into northern Scotland, the Welsh hills, the Lakes of Killarney, or the west coast of Ireland. William Hooker at the University of Glasgow had already demonstrated the popularity of such trips.[63] Watson went along in the summer of 1831 and returned with some of Graham's students to part of the same route in northern Scotland in July, 1832.[64] Watson probably also accompanied Graham in 1829 and 1830. The hypercritical Watson was slow to perceive Graham's limitations, and he dedicated to Graham his *Remarks on the Geographical Distribution of British Plants* (1835).

It was as a collector of plant specimens that Watson sent a note to the *Magazine of Natural History* in October 1829, proposing that someone establish a clearing house for the exchange of natural history specimens. Although his scheme prompted some discussion, it was only implemented in 1836, when the Botanical Society of Edinburgh was founded.[65] Partially crippled though Watson was, his appetite was not satiated by Graham's excursions, so Graham wrote a letter introducing him to William Hooker, and Watson then accompanied Hooker and his class on a field trip up Ben Nevis in autumn, 1829. After Hooker published *The British Flora* (1830) Watson began corresponding with him. On 9 November 1830 Watson requested the names and addresses of individuals who sold dried plants, either British or foreign. He also demonstrated his seriousness as a botanist by questioning the accuracy of one of Hooker's published descriptions:

Is there not a typographical error in your specific character of Erica ciliaris? I have not seen it wild, but the garden specimens have very seldom *4* leaves in a whorl and Lindley writes '3 or 4' ...[66]

Finally, Watson offered to send Hooker specimens from any of 17 species.

---

[62] Wilson & Geikie 1861, 104. Joseph Hooker temporarily replaced Graham during his terminal illness, and his impression of Graham's teaching methods was also unfavorable; Huxley 1918, I, 196–8. For a positive evaluation, see Ransford 1846, 19–20.
[63] Allan 1967, 79–80; Huxley 1918, I, 13–14.
[64] Graham 1832 and 1833. Watson 1832b and 1833a.
[65] Allen 1965. Watson 1830c.
[66] Hooker 1830, 176. Lindley 1829, 174.

Watson's outlook was that of a conscientious student and a discriminating collector. A novice eagerly gathers without much discrimination any objects that strike his fancy, but once he begins to compare his collection with those of others, or attempts to exchange objects with them, he learns the socially determined points of discrimination that can mean the difference between a valuable and an indifferent specimen. A natural object had to be identified properly according to kind and locality. The plants, animals and rocks of Britain were not so well known that a hypercritical collector like Watson could not find gaps in the knowledge. The question of whether a species is native throughout its present range seems simple enough, but it could open the door to some of the great mysteries of plant geography. Watson's comparison of accounts in manuals with plants in the fields might expose mistakes by authors (this, perhaps, being his objective), but it also revealed the variability of species – another unexplained mystery.

Hooker was already arranging to sell American plants at £2 per 100, which Thomas Drummond was collecting. Watson was interested in acquiring a set, and he wrote a number of letters to Hooker acknowledging receipt of them, returning payment, and sometimes complaining about them not being dried properly.

Each spring, Professor Graham offered gold and silver medals for the two best student essays on a specified botanical topic, and in 1831 the topic was the geographical distribution of plants. Watson later recalled that he had 'become a competitor for the medal in consequence of a sneering taunt against the ability of phrenologists ...': It is fortunate that he rose to the challenge, for 'preparation of this essay was probably instrumental in confirming [his] taste for that particular department of Botany'.[67]

When organizing his essay, Watson considered as models Humboldt's 'Prolegomena' and John Lindley's *Introduction to the Natural System of Botany* (1830). Humboldt provided general laws and illustrative examples, while Lindley stated the geographical ranges and degrees of prevalence for each family of plants within a region. Watson followed neither model closely; instead, he divided his essay into two parts, one descriptive, the other dynamic.[68]

His descriptive part begins with a division of the world's flora into six floristic zones according to latitude. This discussion is considerably indebted to Humboldt, and he also drew upon 23 other lists and Floras of various parts of the world. His account of the plants in latitudinal zones is mostly descriptive reporting but is not devoid of theoretical content. His discussions of percentages of different plant families within a zone followed the pattern established by Treviranus, Brown and Humboldt. The significance of these percentages was not fully understood until

---

[67] Watson 1839a, 267.
[68] Watson 1831 MS, 3–4.

after Darwin explained his theory of evolution. The parallel between the latitudinal and altitudinal ranges of plants was already well known, and Watson described it in some detail. An interesting correlation he noted and attributed to temperature was the fact that temperate species had a distribution further north on western than on eastern coasts of continents.[69] Another correlation was that species toward the North Pole tend to be more similar on different sides of the world than are species further south. The striking similarity between the floras of eastern Asia and eastern North America he detected from lists of species from these localities.[70]

Part II, entitled 'Conditions of Vegetation', includes discussions of temperature, moisture, soil and some minor influences. He thought that although moisture is physiologically most important, temperature has the most influence upon the distribution of species.[71] The plausibility of that conclusion can be called into question for aquatic and desert plants, and he did acknowledge the primacy of water when discussing them.[72] He found a diversity of opinions among botanists concerning the importance of soils for plants. Since the same species can grow in different kinds of soils, he decided that soil texture, moisture, temperature and organic remains are more important than its chemical composition.[73] He acknowledged that there were exceptions, such as *Ophrys*, an orchid genus confined in England to chalk soils, and *Erica vagans*, a heath species confined to slates and serpentine soils.[74] He considered as minor environmental factors: shade, protection from cold, parasitism, proximity of other plants, man's influence, and seed transport by man, animals, winds and water currents.[75] He did not discuss these factors, and therefore slighted dynamical aspects of plant geography. These omissions are excusable in an otherwise impressive student essay (which won the gold medal). However, he never fully made up for his neglect of plant dispersal in later studies.

When Watson left Edinburgh without the MD degree, he was more committed to phrenology than to botany. However, when he later committed himself to botany, the lack of a degree was still not a serious problem. Two contemporaries who were later prominent naturalists, Charles Darwin and Edward Forbes, also left Edinburgh without medical degrees. All three had prospects of independent wealth, though in Forbes's case the prospect was short-lived.[76] (Darwin later earned a BA degree at

---

[69] Watson 1831 MS, 70.
[70] Watson 1831 MS, 73–4.
[71] Watson 1831 MS, 77–8.
[72] Watson 1831 MS, 85–6.
[73] Watson 1831 MS, 90–4 & 98.
[74] Watson 1831 MS, 94.
[75] Watson 1831 MS, 99–100.
[76] Egerton 1972. David E. Allen adds a fourth name to this list, from an earlier generation: John George Children (1777–1852); see Allen 1985, 8.

Cambridge because his father envisioned a career for him as a clergyman.) Nor was a degree necessary for teaching botany in most British universities. William Hooker was appointed Professor of Botany at Glasgow without ever having gone to college, and Forbes, without a degree, was appointed to chairs at two different universities.

Despite plant geography being a less well organized science than phrenology, Watson wrote his first book on it: *Outlines of the Geographical Distribution of British Plants; belonging to the Division of Vasculares or Cotyledons* (1832). This was possible because of his acquaintance since childhood with British plants, along with the preparation which he had made before writing his prize essay. Humboldt's plea for a new science of plant geography encouraged him to work in this field. *Outlines* is divided into two parts, the first being a general discussion, and the second providing a brief indication of the habitation, topographic range and worldwide distribution for vascular species found in the British Isles. He borrowed this organization from Wahlenberg. Although research for the prize essay helped prepare him for the task, *Outlines* was organized differently, drew upon new sources, and exhibited firmer control over subject matter because of a more restricted geographical scope.

Watson doubted that 'a complete panorama may ultimately be constructed' from local studies: 'the respective Essays of Messrs Winch and MacGillivray are good pictures, but they would not unite'. Watson realized that the advancement of science is a group endeavor, and he sought to impose a uniform approach under his own guidance.[77] He began by asking how plants got where they are, and his survey of possible answers shows he was uncertain about them. He cited Winch's estimate that almost fifty species had spread into the Northumberland and Durham hills from the ballast of ships. Watson pointed out that introduced species persist only when they encounter congenial climate and soil; as proof, he cited the seeds of American tropical plants which fail to survive when deposited on British shores by the Gulf Stream.[78] From his prize essay he borrowed a discussion on the factors most important for influencing the distribution of plants: temperature, moisture, soil and situation. The section on climate and physical aspects of Britain presented data on temperatures, including the mean annual, winter, spring, summer and autumn temperatures, and the hottest and coldest months for Penzance, London, Edinburgh, Aberdeen, and Kendal.[79] He also gave data on annual rainfall and elevation of about 530 places in England and Scotland. His 'Outline of the Topographical Distribution of British Plants' divided British vegetation into three regions, each of which he subdivided into two zones, as shown in Table 2.1.

---

[77] Watson 1832a, vi–vii.
[78] Watson 1832a, 1–4.
[79] Watson 1832a, 14–17.

Table 2.1 Watson's regions of British vegetation

I. Woody Region	1. Agricultural zone
	2. Upland zone
II. Barren Region	3. Moorland zone
	4. Subalpine zone
III. Mossy Region	5. Alpine zone
	6. Snowy zone

Each zone contained indicator species which were absent from the zones above and below:

1. Agricultural zone ends where cultivation of wheat ceases
2. Upland zone ends where cultivation of *Corylus avellana* ceases
3. Moorland zone ends where cultivation of *Carex rigida* begins
4. Subalpine zone ends where cultivation of *Calluna vulgaris* ceases
5. Alpine zone ends where cultivation of *Empetrum nigrum* ceases
6. Snowy zone ends where cultivation of land terminates.[80]

For each zone he indicated the geographical range and species. Later on, as Watson's knowledge increased, he modified the limits of these zones, their names and indicator species, but he always retained six of them.[81]

This last detail is important for a dispute with Edward Forbes that began in 1845. Watson's zones were useful, and his works were highly regarded by other botanists; nevertheless, his system was one of convenience. He had not 'discovered' some absolute reality which could only be described his way. Later authors, such as A.G. Tansley, changed over to a classification that has less emphasis on latitude and altitude, and more on types of vegetation: Woodlands, Grasslands, Hydroseres, Heath & Moor, Mountain Vegetation, Maritime and Submaritime.[82]

Although Watson never used his information to test hypotheses about distributions, he did use Humboldtian correlations in an attempt to generate new laws with his data. The general information in Part I is followed in Part II with the distribution of British species arranged by family.

---

[80] Watson 1832a, 35.
[81] Watson 1835a, 57–72, esp. 57–8; 1847–59, I, 35–55, esp. 40–1.
[82] Tansley 1911 & 1939. On Arthur G. Tansley (1871–1955), see Desmond 1977, 598–9. On early parallels between the thinking of Watson and Forbes, see Rehbock 1983, 125–8.

In the summer of 1832 he went on a field trip into northern Scotland with some of Graham's students, who soon left him because he was more interested in measuring elevations with an Adie sympiesometer[83] and temperatures with a Fahrenheit thermometer than he was in collecting plants. He attempted to relate his measurements on both environmental factors to the distributions of species. He found the same species at different elevations on different mountains, and decided: 'Absolute altitude is of little importance in the geography of plants, [and therefore] my attention was for the most part limited to the observation of their relative height in regard to each other.'[84] He listed the upper or lower limits for several species at four locations; two examples (expressed in feet) are shown in Table 2.2.

Table 2.2 Watson's delineation of upper and lower limits for two species at four locations

	Upper limit on			
Species:	Clova	Braemar	Fort William	Tongue
Erica cinerea	2400	2200	2100	1730
Ulex europaeus	1500	1350	280	350

Although he did not explain why he thought these variations occurred, he gave relevant information. One important factor was 'situation', by which he evidently meant both angle and direction of slope: 'The influence of situation is well exemplified by the fact that *Empetrum nigrum*, under the steep snow rocks on the northern side of Ben Nevis, fails 600 feet below its height on the western side.'[85] Another clue is his explanation of why wheat could not be cultivated on any of these locations: 'Braemar is too high; Fort William is too wet; Glen Clova, a narrow valley shaded from the sun by high hills; and Tongue, exposed by a north sea, and with high ground to the south'.[86] Neither his charts of upper and lower limits of plants nor those of temperatures specified on which side of a mountain the measurements were made. Furthermore, he failed to specify the difference in amount of light the plants received in different situations on the mountains even though he listed this as a significant factor. He did give, as Humboldt had urged, the mean temperature for the

---

[83] This instrument was used for several decades as a marine barometer; Adie 1819; Middleton 1969, 38.
[84] Watson 1832b, 357.
[85] Watson 1832b, 358.
[86] Watson 1832b, 361.

hottest month, 16 July to 17 August, which was 52°F at 2000 feet elevation.[87] He also averaged his measurements to obtain a general temperature gradient, from 1000 to 4000 feet elevation. He listed the species he collected with their locations and elevations. Finally, he prepared a chart showing the percentage of species from each plant family found at different elevations. Watson's two-part report resembled Humboldt's *Essai* in failing to establish a definite connection between the plants and any environmental factor except elevation. Yet Watson's own experience was that elevation alone told little. Perhaps his tendencies toward skepticism and criticism were held in abeyance because he was just beginning his researches.

In 1832 Watson gave up on a medical career, but he made no decision between being a phrenologist and a botanist. He decided he wanted to go and study American plants, and on his behalf Hooker contacted Sir John Richardson, a naturalist who had been on Franklin's second Arctic expedition (1825–27). Richardson replied that he could arrange such a trip. Faced with the real possibility, however, Watson found several excuses, which he explained to Hooker (20–22 December): he doubted he could find suitable companions, Mexico was more interesting than Canada, and so on.

---

[87] Watson 1833a, 318. See also Browne 1983, 66–7.

PART II
THE LIFE OF A GENTLEMAN SCIENTIST

Chapter 3

Relationships and Social Perspectives, 1833–59

> I spent last week at my Brother Hewett's new house about 15 miles from London & ... he has got a beautiful garden which he takes care of almost entirely himself.
>
> Julia Watson Gorton to E. Dawson, 19 June 1834

Private Life

Watson claimed, late in life, that he had left medical school because 'as he was going up for his examination his health broke down, and he was not able to go up for his degree'.[1] He did not state whether this breakdown was physical or mental. Presumably, it occurred during the first half of 1832. He spent the year 1832 in Edinburgh, active in both phrenology and botany. When he left in January 1833, he went to live with his sister, Louisa Judith, and her husband, Captain Gilbert Wakefield, near Barnstaple, Devon.

A remote Devonshire town was in striking contrast to the cosmopolitan environments of Manchester, Liverpool and Edinburgh where he had spent the previous decade. This is how our young phrenologist explained the fringes of civilization to George Combe (6 March 1833):

> Here the good people are so ignorant, so self-satisfied, so antiquated in their opinions, that quitting Edinburgh for Barnstaple is like emigrating from Britain to China, or stepping back 50 years in the march of time & of intellect. You may fancy a society made up of retired officers, and families of moderate independencies, who hold trade in contempt, yet envy its wealth; who have no occupation save hunting, shooting, fishing, card-playing, quadrilling, walking, & backbiting each other. Phrenology is quite unknown, except as something that Dr. Budd believes! I do not however ascertain that the said Dr. Budd *knows* it, nor have [I] met with him in person. Geology is taught at the Ladies Seminary, but

---

[1] Baker 1883, xi.

specimens are deemed quite unessential, the *names* in books being all that is requisite to be learned; to what *objects* the names apply is unimportant.

Without the benefit of Combe's Sunday breakfast conversation, Watson lamented, 'I shall become more & more rusty & moody.' Still, Watson was also a botanist, and any provincial town held some interest from that standpoint. To Hooker's inquiry he replied on 23 January that 'No snow has fallen, & there are some 30 or 40 wild flowers still in bloom, chiefly autumnal ones.'

Nine months later he wrote to Hooker that he had no wish to renew the lease on the house which he had rented jointly with Captain Wakefield for a year, and that he was leaving that day (15 September) to examine a house for sale at Kingston in Surrey. Not long afterwards, he bought a house which he occupied for the rest of his life. John G. Baker, who visited him there several decades later, described it as a 'small house (which he afterwards enlarged) with a pleasant garden and orchard attached to it, a short distance south of the village of Thames Ditton, a stone's-throw from the point where the branch to Hampton Court leaves the main London and South Western [railroad] line'.[2] According to one historian, Watson's house was named Fern Cottage, though Watson did not include that name as part of his return address on letters. 'Fern Cottage was in Manor-road North, but it was demolished some years ago ... He also owned two houses in Claygate Late known as North and South Cottages, which [in 1970] still stand.'[3]

When, on 21 April 1834, he invited Nathaniel Winch to visit him, Watson warned that his cottage was 'yet in a very rough state inside & out'. He soon fixed it up. Watson explained to Combe (29 August) that, 'though little more than a dozen miles from the metropolis, unsocial habits place me "out of the world",' and he also emphasized to Hooker (27 August) his love of solitude. Yet, London was as important to him as the countryside. He appreciated the resources provided by the Linnean Society of London, which he joined in 1834. Despite his imagined 'lack of harmonious feelings' with his sisters,[4] three of them lived in London, and he maintained a relationship with them and their families.

He inherited enough wealth to live comfortably, if not lavishly. His 'extravagant' brothers[5] both died young and would not have had much, if any, wealth to pass along, and whatever they had would not have been left to their frugal, disapproving

---

[2] Baker 1883, xi. Watson also described both his Barnstaple experience and his new situation in a letter to W.C. Trevelyan, postmarked 28 March 1834.
[3] Mercer 1970, 59. See also Mercer 1971.
[4] Watson to Combe, 22 November 1834.
[5] Watson 1839c, 265, used this word to describe them. The father, Holland Watson, left a complex will, but it seems that his eldest son was his main beneficiary and that the two younger sons received less than some of the daughters.

older brother. When he applied for a position as a botany professor in Ireland in 1846, Watson understood that the salary would be £300 per year, and that was a major aspect of his interest in the position – not because he needed the money, but because it would enable him to expand his plans for scientific publication. He explained his financial situation to Combe (25 August 1846):

> There are advantages to be gained by removal to Ireland, though, if richer, I would keep to the vicinity of London. My income is enough for comfort, where display is not cared for; and even there is something spared for others; from a tenth to a fifth is drawn from me by the appeals for aid from others, tho' I am often compelled to refuse appeals; for, if all were acceded to, I should not have a shilling a year over. But I wish to publish results of much labour and time, which will never repay the cost of publication, and an addition to my income would be one of the advantages, on this account.

Anyone who reads the letters between Watson and the Combe brothers over the finances of *The Phrenological Journal* might wonder whether Watson was exaggerating the extent of his charitable contributions. However, George Combe was unlikely to have been skeptical, since over the years he had witnessed Watson's charity on a number of occasions, beginning with assistance to a desperate Edinburgh sculptor in 1832.[6]

Watson liked solitude, but was not a recluse. He disliked most meetings, though he attended some. Many naturalists, including William Hooker and Charles Darwin, visited him, as did friends and relatives. He never married, but there is no evidence of either a homosexual tendency or of sexual indifference toward women. While languishing in Barnstaple, he frequently thought back to his enjoyable visits to the Combe household, where his scientific judgment was respected, the conversation sophisticated, and Marion Cox poured tea and sometimes teased him. When writing to George Combe on 6 May 1833 he asked to be remembered to her. Her response, relayed by Combe (16 May 1833), conveyed more than mere politeness: she sent her kind remembrances and wished he could come again for breakfast, but was sure his boredom would soon be relieved by the beauties of Keswick. Marion Cox was about a year older than Watson, and their breakfast-table relationship seems to be as close as he ever came to finding a woman to marry. In 1815, Robert Cox, brother-in-law to the Combe brothers, died at age 42, leaving his wife, Ann Combe Cox, with three sons and two daughters to raise. George and Andrew Combe were very close to her

---

[6] Watson to Combe: on contributions to the sculptor, 30 June and 27 August 1832; other charitable donations, 6 May and 22 August 1833, 29 August 1834, 25 August 1846, 7 October 1853, 26 December 1854.

and her children, and must have made substantial contributions to raising them. Robina married, but Marion became the indispensable manager of the household of her two uncles. It was in that context that Watson had known her.

When George Combe wrote announcing his engagement to be married, Watson was shocked and protested! To upset that idyllic scene by importing a new hostess must be unphrenological! Watson knew that at an early age George had decided not to marry, because of a 'scrofulous' skin infection, which might be transmitted to wife and/or children. The sickness in question was tuberculosis, which was indeed a peril to the Combe family.[7]

Combe's explanation of his decision (16 May 1833) shows that Watson's approval was important:

> I expect to be married in September to Cecelia, youngest daughter of the celebrated Mrs Siddons ... She is 39 and I am verging towards 45; so that my object is to have a companion capable of yielding me affection and also moral and intellectual sympathy ... You may perhaps doubt if I am obeying the [natural] laws, in marrying at all, with my constitution; but I consulted Dr Spurzheim in 1828, telling him *all* my constitutional history, & he said he thought I might lawfully marry, providing the lady was not too young ... I have delayed till this age, and have now chosen a partner whose years will prevent the possibility of *many* children. If they come at all, there will be few to suffer, in case I have been wrongly advised ... I write this freely to you, although confidentially, because you are a philosopher and can judge of my conduct.

Watson's response was confused, and he groped for a phrenological justification for his feelings (22 August 1833):

> ... what appeared to me to render this state a matter of very great improbability in your case, was the finding a partner possessed of all the requisites according to your own views of matrimony ... You allude in your letter to the risk of not acting in conformity with the natural laws. This emboldens me to make some comment thereon. I think you are perfectly right in consulting your own wishes, as an individual person, in this instance; & yet do not think you acted rigidly in accordance with your written opinion. Perhaps the non-conformity may be only in my fancy, not in your acts.

That Watson's responses were based upon feelings about himself as well as Combe is also expressed in this letter:

---

[7] Gibbon 1878, I, 284–5.

Figure 3 Hewett Cottrell Watson (1839), said to be drawn by Haghe, but bearing the initials 'R.D.D.'

For my own part, I have pretty nearly determined against matrimony, but bachelorship to a person of excitable, though not durable, feelings is attended with disadvantages easily conceived, & may cause as much evil as would marriage.

While George Combe no longer needed Marion Cox as a housekeeper, Andrew still did. Andrew already exhibited the symptoms of pulmonary tuberculosis by autumn 1831. He managed to stretch out his life until 1847 by rests and winter vacations in Italy, the Madeira Islands and other warm places. Marion was his dependable nurse, at home and abroad.

Watson had an adequate excuse for never pursuing Marion romantically – it would have seemed selfish to try to take her away from the Combe brothers. But other young men in his shoes might have thought it worth asking if she would not prefer a marriage to housekeeping, and let the brothers find someone else. He was inhibited by the fact that in some sense she seemed 'taken', and also most likely by a degree of shyness toward eligible women of his own class which was tied to his physical disability, even if that was only a rationalization.

Combe did not debate Watson on the merits of marriage; he merely mentioned that he was happier after marriage than before. Watson tried to understand (22 November 1834): 'It is not likely that your preference should be directed toward any one whose mental qualities would be discordant; & veneration would tend to remove acquired differences of thought ...'. Watson continued to turn over in his mind his own conflicting opinions, and five years after their initial correspondence on the subject he notified Combe (10 June 1838) that he had changed his evaluation of marriage. That was a hint for Combe to ask for the new evaluation and he obliged (7 July). Watson's reply (21 July 1838) was partly a neurotic attempt to repair a non-existing damage to their relationship brought on by the earlier correspondence and partly an attempt to justify phrenologically the success of the Combe marriage.

On a more personal note, Watson continued:

... in your letter to Mr. Jones, you made a pledge on my part somewhat too hazardous, in writing 'I can safely pledge my honour that Mary would be as safe in his hands as in her mother's.' The circumstances of you having made such a pledge would, no doubt in itself increase the safety of making it; but most men are ready to act towards females of humble rank, without any regard to ties of honour, truth, or self-forbearance, when their amative feelings are concerned.

Although Watson rarely saw himself in the category of 'most men', if he exempted himself from being susceptible to the above situation there would have been no need to write that paragraph. What these comments indicate is that he was hiring Mary Jones as a maid, and that maybe he regretted Combe's assurance to her father that

Watson would not exploit her sexually. If he did hire her, she would have been supervised by his housekeeper, Grace Eastmond. In 1842 he wrote a will that shows more concern for Eastmond than for anyone else (she being named in it five times, though he spelled her surname 'Eastman'). Had he died then, he would have left her a substantial income.[8] He referred to her once to Combe (8 January 1855) after Combe sent him a travelog he had published (*Notes on a Visit to Germany in 1854*). Watson 'read [it] with gratification as showing the actual condition of peoples, places, etc.' He then:

> passed [it] into the kitchen, and recommended to be read there. But in saying this I should properly add also, that one of the servants has lived with me over twenty years, and was brought up by an estimable and judicious mother; consequently, much better able to understand and appreciate your letters, than would be expected in the case of a domestic in a small house.

Is Watson describing here not only his housekeeper, but also his mistress? A botanist and gossip, Thomas Bruges Flower, who after Watson's death claimed to have been his friend, stated that Watson had a child out of wedlock by his housekeeper.[9] John G. Baker reported that Watson 'was fortunate in that the same housekeeper who came to him when he first settled down in a house of his own remained with him for the whole forty-six [closer to 48] years, and nursed him through his last tedious illnesses'.[10] There is an otherwise unidentified 'protegee' who received a £200 trust fund in his final will, Alice Ada Parfett, who might have been his child. One difficulty is that she had a sister, Amelia, and a brother, Donald, whom Watson called 'my friends'.[11] Another difficulty is that the ages seem wrong. The language of his will written in 1878 (which retains Watson's misspelling of Eastmond) seems appropriate for a girl in her teens, which means that Watson would have fathered her in his fifties; even less plausibly, Grace, the putative mother, was probably older than he, for in 1872 he referred to her as 'my old Housekeeper' (letter to J. Hooker, 26 Sepember). If they had had a child, it would probably have been born one or two decades earlier than Alice was. Within the confines of available evidence, there is not sufficient reason to accept Flower's claim.

---

[8] I am indebted to David E. Allen for finding this will at the Royal Botanic Gardens, Kew, and providing a copy to me.
[9] Thomas Bruges Flower (1817–99), as related to William Bowles Barrett (1833–1915) in 1882. MS notebook at Weymouth Public Library, see B. 22. This information is supplied by David E. Allen. On Flower and Barrett, see Desmond 1977, 227 and 41.
[10] Baker 1883, ix.
[11] Watson, MS 1878, l. In this hand written copy of Watson's will, the name appears to read 'Parfelt', but David E. Allen says it should read 'Parfett'.

When buying his house, Watson would have been interested in gardening. A garden provided the happy opportunity to unite his scientific, esthetic and creative feelings. Many of his letters mention efforts to obtain seeds of species from particular localities to grow for his research. He expressed to Combe on 5 March 1841 what must have been a frequent feeling:

I have for many days been so completely a working gardener, that my fingers are now almost too deadened in sensation to guide a pen. I am much tempted to think that hard work in the open air is the happiest life ...

Once he was tempted, as he wrote to William Hooker (3 March 1842), to raise all the species of Britain, only to realize that such a project would require two gardeners, which was more than he could afford. He could, however, as he informed Hooker (30 June 1844), send his gardener to Kew 'to look through the gardens there, by way of gaining a few ideas'. After Eastmond had been his housekeeper for four decades, he rewarded her with a trip to Kew, for she shared his interest in flowers (to J. Hooker 26 Sepember, 1872).

Watson did not become so absorbed in his domestic routine that he neglected friends. Whenever Andrew Combe's and Marion Cox's travels took them through London, they stopped to see her brother, Dr Abram Cox, and also Watson. A letter from Watson to George Combe (4 October 1846) indicates that she still enjoyed teasing Watson, and that he still responded by trying to impress her. In reference to printed recommendations for a chair of botany, Watson asked: 'Pray transfer this copy to Rutland St. and request Miss Cox to read the certificate of [George Edgar] Dennes (No. 7). When in Kingston, Miss C. seemed curious to ascertain how causality operated in Botany.' After Andrew's death she assisted her uncle George to write Andrew's biography. Combe asked Watson to read page proofs of the book and he gladly agreed (22 February 1849). Thus Watson and Marion Cox cooperated on this, her final endeavor before she, too, died.[12] Watson, an effusive critic, lacked experience in expressing positive feelings. He therefore resorted to the English capacity for understatement when he wrote to Combe after Marion's death (19 April 1850): 'With much pain I learn from Robert Cox the state of his sister. Towards her there has been much stronger regard than you probably supposed the case, on my part. She was esteemed, and liked because esteemed.'

---

[12] Watson to Combe, 27 February, 7, 12 and 28 March and 18 April 1849. Watson had expressed his great esteem for Andrew to G. Combe, 7 January 1843. G. Combe 1850. Gibbon 1878, II, chap. 8. Marion Cox died on 22 April 1850, after three weeks of illness; Gibbon 1878, II, 288–9.

Only one of Watson's acquaintances seems to have been active, like himself, in both botany and phrenology: Sir Walter Calverley Trevelyan.[13] In 1834 there was a chance that he would come for a visit, and on 27 June Watson sent him a diagram of how to find his place, but Watson's letter to him on 15 December does not mention that he ever came. Despite their cordial interactions in both fields for a number of years, Watson confessed (8 October 1838) to a reclusive timidity when Trevelyan invited him for a visit in Newcastle upon Tyne:

> For the wish you so kindly express, that I should visit you, if going to Edinburgh next spring, I am greatly obliged, & to myself it would be gratifying to do so. But I am by nature reserved almost to timidity, & so little accustomed to society, that I am scarcely fit for it; & hence have pretty constantly the idea that others must think 'what a dull & stupid person H.W. is, I wish he was gone'.

Whether or not Watson stayed with the Trevelyans, he did visit them, for in the next year he wrote (30 Sepember 1839) asking what news they had from Miss DeCourcy, who had been visiting from Italy while he was also at their home.

Family Life

Watson's comments to others about members of his family were usually negative.[14] When applying for the chair of botany at King's College, London, in 1842, and concerned over his political and religious acceptability to the college, he complained to W. Hooker (n.d., no. 246):

> My father was many years an active county magistrate, known for extreme Toryism, and as I could not by any means implicitly adopt his ultra views, I unavoidably became a sort of black sheep among the white, Tory associates of the family; the points of agreement going for nothing, while those of discordance were magnified.

On another occasion he lamented to Combe (2 August 1848) that none of his sisters measured up to his mother in running their households:

> The capacity to do [housewifely] acts was usual in matrons of families half a century ago, with the exceptions of the two [economic] extremes of society.

---

[13] On Trevelyan (1797–1879), see Desmond 1977, 618 and Watson's letters to him.
[14] Watson to Combe, 22 November 1834. Flower to Barrett in 1882, Weymouth Public Library MS notebook, p. 22.

Nowadays, much less usual, I believe, to find the same practical training to household affairs, unless in quiet country places.

For instance, my mother was well skilled in such matters, although brought up as a lady, when social distinctions were more decided; her father probably enjoyed an income of some thousands a year. Of her seven daughters, perhaps not one is so competent in household affairs, although skill of that kind should have seemed more necessary to them, as their individual fortunes were equal only to one-fifth of that of their mother. It is of course the changing habits of society which render household skill of the homely kind less needful now.

However, because of his propensity for negative assessments, Watson likely exaggerated the depth of the disapprovals on both sides. His family loyalty and affection is illustrated in his comments to George Combe about works which he and Andrew had published on education (29 August 1834):

Your brother's work particularly pleased me, as I had often wished for something of the kind to put into the hands of married sisters, bringing up families, but had never met with any work sufficiently simple & at the same time, sufficiently sound.[15]

One sister whom he attempted to improve was Caroline. Watson was as close, or closer, to her and John Horatio Lloyd as to any other of his sisters and their husbands. Besides being a brother-in-law, John was Hewett's first cousin, and conservative though he was, he surely expressed genuine esteem in the letter which he wrote in 1846 supporting Watson's application for a chair of botany. It is addressed to Watson, but meant for transmittal to the university:

if an earnest pursuit of truth, extensive and accurate knowledge, a diligent cultivation of science, not in one department only, though, perhaps, more especially in that of botany, in its widest application, combined with a simple, perspicuous and methodical style in speaking as well as writing, constitute peculiar qualifications for a Professorship, you may justly assert your claim on these grounds. I am sure also that, if appointed, you will conscientiously fulfil the duties of the office, and endeavor to carry out, honestly and faithfully, the design of the Institution.[16]

The persistence of this esteem is indicated by the visit of several days which Caroline and her daughter Louisa Mary paid to Watson: they are unlikely to have

---

[15] The books alluded to were A. Combe 1834 and G. Combe 1833.
[16] Quoted in Watson 1846a, 39.

stayed so long if they found his company disagreeable. They arrived shortly after George Combe published the biography of his brother, and this provided an occasion for Watson to view, and report to Combe (19 April 1850) the disappointing results of his efforts to improve these ladies' minds:

> I read portions from the volume one evening to my sister and her daughter, ages 47 and 16 ... My sister had once seen Dr. Combe, who dined with me at her house, many years ago. She attended to the portions read, and remarked on them. But during the two days after, which they spent here, I did not observe either of them take up the book from the table. The young lady preferred a novel; her mamma looking at Chamber's Miscellany, Soyer's Cookery books, etc. – This sister never seemed to take any interest in phrenology or physiology.

Two years later he assumed a typical uncle's role – trying to protect the fair maiden from an unscrupulous lover. The suitor under suspicion was William Napier (1821–76), younger brother of Lord Francis Napier. Watson observed him long enough to reach a phrenological judgment: 'Large knowing organs and language, moderate reflection, coronal surface less than I like, and aution defective'. He transmitted this assessment to Combe (11 November 1852) with a request for information about Napier, and he explained why he asked:

> He and a niece of mine have been thrown much together on the Continent in the past summer. Attachment, explanation, engagement, the natural consequences, – but much regretted by myself, on various grounds, among them the important one that he has no fortune and no profession, merely hoping for an appointment through family influence; while she is a charming girl; only 18 (he 31), accustomed to the comforts that a paternal income of 3 or £4,OOO a year will afford, but most of it to die with her father, now aged 55 or upwards [J.H. Lloyd died 18 July 1884, aged 85], and she is one in five sons and daughters. Promising prospect, between an indefinite engagement at first, and poor widowhood afterwards!

Watson blamed Napier for 'seeking to win the affection of an inexperienced girl, under the circumstances; she having no knowledge of his straitened means'. Watson understood that 'his family thinks the engagement unwise and objectionable and so do we ...'.

Combe confirmed Watson's phrenological judgment: there was indeed evidence of an irresponsible past. Watson confronted his sister and Napier with the evidence. To his surprise (he reported back to Combe, 17 November 1852):

> ... she already knew of the McA. scandal, – first from another quarter, and then from W.N. himself, whose version is less unfavorable to himself, and more unfavorable to the husband, than was the account as penned by Dr. C[ox?]. I regretted to find my sister, the mere advocate of W.N., willing to rely implicitly on his assertions, and disinclined to investigate. With this blind infatuation in the mother of La Fiancée, and apparently the father pretty much in the same way, I become only an unauthorized intermeddler.

Crestfallen, Watson backed away, but a few days later he reported (25 November) that further negative information from other sources had finally caused Louisa Mary's parents to ask the Napier family directly about William's character. Their disclosures led to termination of the engagement. Watson marveled at the slowness with which cousin John had responded to his daughter's peril:

> Is it not strange that persons who believe in phrenology, and are to some extent able to apply it cranioscopically and analytically, should yet so little use their knowledge in the most important affairs? My brother-in-law is a man of good and trained intellect, far above average, and pretty well conversant with phrenology. He received W.N. as the suitor of his daughter, and had thus full opportunity for observing cranioscopical indications, and trying their truth by conversation, etc. And yet, the need for inquiry into past conduct never seemed to occur to him, until forced upon him by others.

However, even the Napier family confession did not end the story. Watson had probably underestimated the size of Napier's amative faculty, for he regretfully informed Combe on 24 May 1853:

> my niece is still under a conditional engagement with the Hon. W.N., having the full sanction of her mother, and the not very willing assent of her father thereto.

Since Louisa Mary's parents stipulated that William must have an annual income before marriage of £1000, Watson thought there was little chance for the marriage ever taking place. A closer look (presumably still without feeling the head with his hands) led Watson to a slight revision of his initial diagnosis: 'W.N. is not premeditatingly vicious; but a man of present impulse, without enough of caution, or conscience, or causation, to restrain and guide.'

In the event, William Napier married Louisa Mary Lloyd on 3 May 1854. After having observed Napier's married behavior, on 24 October 1855 Watson reported, with satisfaction, to Combe:

His conduct has been very much above what the report might reasonably have led us to anticipate. Evidently, his own family holds him in much higher respect than they did three or four years ago. He is thoroughly a Napier, in his deficiencies and gifts, and up to the average of the family. He has needed, and perhaps still requires, some judicious Mentor to guide and instruct him. He found this Mentor not in any single person, but in the aggregate family with which he connected himself.

Although Napier came from a family of means, Watson was probably right about him needing a mentor, for his father had died when he was only 13. The consequences of this loss were difficult to detect cranioscopically. The marriage proved successful, at least to the extent that William and Louisa Mary had eight children, whom Watson was to remember in his last will.[17]

In Christmas greetings to George Combe in 1854, Watson complained about the burdens of family and then admitted there were compensations:

... 1854 has been to me the most busy, and perhaps the most disagreeably busy, year for some twenty years past. Heavy losses, incurred by lending money to embarrassed friends, have been added to various other more vexing, though intrinsically minor causes of annoyance. And large demands have been made on my time, to perform troublesome, and often unpleasant work for other people.

However, the retrospect of such occupation of time usually becomes satisfactory; and it rarely fails to be rendered so, by regarding it as repayment to D.E.F. for benefits derived by myself from the time and labours of A.B.C., etc., whether predecessors or contemporaries.

---

[17] Watson, MS 1878, 2.

Chapter 4

Phrenological Struggles, 1833–40

> Ages may perhaps elapse before Phrenology shall be sufficiently advanced towards its maturity, as a science ...
> Watson 1838a

Watson conducted few phrenological investigations after he left Edinburgh, but he nevertheless remained an active phrenological author. Much of what he wrote was belligerent in tone.

In 1835 George Hancock submitted a 'Letter on the Functions of the Organs of Comparison and Wit' to *The Phrenological Journal*, in which he commented on views of George Combe and Scott but ignored Watson's.[1] Watson quickly responded with an angry critique, and called attention to his earlier papers. Realizing that his tone was harsh, he ended by assuring Hancock that no personal attack was intended, 'although the necessary intrusion of some egotism in my first paragraph might possibly suggest such an idea'.[2] Hancock's reply shows considerable restraint, and a tactful editor might have ended the matter there, but Watson was allowed to have the last word.[3] George Combe clearly felt that Watson's reasoning was incisive, for he quoted extensively from Watson's article on Wit in the fourth edition of *System of Phrenology*.[4]

In 1838 another naive phrenologist, J.R. Rumball, fell into the same trap of submitting a letter to *The Phrenological Journal* on the 'organ' of Wit without being familiar with the prior literature. This time, however, Watson was the journal's anonymous editor. Rather than return the letter to its author, he published it preceded by a sharp critique that praised the earlier analysis of Mr Hewett Watson.[5]

Although in earlier years Watson had some interest in ornithology,[6] his reason for discussing 'Why do birds sing?'[7] seems best explained by a paradigm that focusses

---

[1] Hancock 1835.
[2] Watson 1835e. The MS of these comments is preserved in the Watson letters to G. Combe, 2 September 1835.
[3] Hancock 1836. Watson 1836e.
[4] Combe 1836a, I, 417–22.
[5] Rumball 1838, Watson 1838b. On Rumball, see Cooter 1984, 294.
[6] Watson 1839c, 267.
[7] Watson 1837f.

an investigator's attention upon particular questions.[8] According to Gall's theory, animals with a conspicuous, faculty will have a correspondingly large brain 'organ' devoted to it. Comparative anatomy could provide a correspondence check with the human brain (where that organ might be less conspicuous and so less easily detected). Phrenologists, therefore, were interested in studying the faculty of Tune in birds. Watson's contribution was speculative and inconclusive.

In only one article did Watson provide statistical data: 'On the Relation between Cerebral Development and the Tendency to Particular Pursuits; – and on the Heads of Botanists'. His avowed aim was to contribute to the development of phrenology as an aid in choosing a profession. He 'discovered' that botanists who confined themselves to the relatively simple tasks of collecting, identifying, and classifying plants had well-developed organs for *Individuality, Locality, Size, Number* and *Language*. The few who went further and studied physiology had large organs for *Causality* and *Wit*. Thirdly, philosophical botanists had large organs for *Comparison* and *Ideality*. Several botanists to whom he alluded had particular brain organs which were too small for the projects they undertook, and their published works seemed correspondingly flawed. Later, when he published an autobiographical sketch (ostensibly written by another), his own cranial characteristics indicated that he belonged in the third of his categories for botanists. With data from only twelve botanists' heads, he confidently announced that 'the average of botanical heads is smaller than will be found in those of several other sciences, as Geology, Moral Philosophy, or Political Economy'.[9] A skeptic might have asked for the data on these other kinds of scientists.

By 1836 George Combe had made enough money from practicing law, supervising the family brewery and from publishing books to be able to devote his full time to phrenology. The Professorship of Logic at the University of Edinburgh became vacant and he stood for the position. To support his candidacy, he published a booklet of testimonials from clergy, physicians and others, including Watson. However, the opposition had more influence, and the chair went to his old nemesis, Sir William Hamilton.[10] Perhaps that defeat emboldened William Scott, a former president of the Edinburgh Phrenological Society, who published *The Harmony of Phrenology with Scripture: Shewn in a Refutation of the Philosophical Errors Contained in Mr Combe's 'Constitution of Man'* (1836, 332 pp.). Such an earnest challenge proved irresistable to Watson, though he complained to Combe (4 September 1836): 'Had you written "this is" instead of "the Creator intended this," Scott's remarks would have been met with ease, – nay, many of them would not have

---

[8] Kuhn 1970, 37.
[9] Watson 1833c, 102.
[10] Gibbon 1878, I, 331–3. Watson's letter is in Combe 1836b, 6–7.

been made.' Since both Combe and Scott were phrenologists, religious sensibilities were Scott's main concern in his attack. Watson, however, did not make religion the focus of his *Examination of Mr Scott's Attack upon Mr Combe's 'Constitution of Man'* (38 pp.). Rather, he focussed primarily upon Scott's careless procedures. Later, Combe's biographer stated: 'Mr Watson, by collating a single chapter of the Harmony with the text of the Constitution, exposed so many misquotations and perversions of the sentences that, although a second edition and a "people's" edition of the attack were announced, they never appeared.'[11]

Watson's major exertion in 1836 was an interesting little book, *Statistics of Phrenology; being a Sketch of the Progress and Present State of that Science in the British Isles*. It was more of a geographical survey than a statistical treatise. He was experienced in botanical geography, and that background helped orient this book. Its title, however, reflects the fact that statistical thinking was becoming something of a fad. For example, Watson's childhood acquaintance, Rev. Dr Edward Stanley, around 1836 decided that parish clergy – who were not always very busy – should collect data in their parishes which could then be made available for general use regionally and nationally.[12] Watson expected his book to help phrenology grow. Phrenologists had 'been accustomed to meet with opponents more frequently than with friends',[13] and his book should increase their awareness of each other and further national cohesiveness. Writing it was not easy because of problems in collecting data. He hoped to collect his information mainly from responses to a questionnaire which he inserted in the June 1836 issue of *The Phrenological Journal*. However, he complained to Combe (23 July 1836): 'It has proved six times more difficult to get any aid and information from phrenologists than to obtain the same co-operation in botanical matters, from brother botanists. I fear there are yet few really generous phrenologists ...'. Where his questionnaire yielded no local response, he would write directly for information. For example, he requested that Botany Professor John Henslow check on the Cambridge Phrenological Society. (He replied: 'I rather think it no longer exists.'[14])

Statistics of Phrenology begins with a historical survey of conflicts in Britain. Phrenologists' replies to critics seemed adequate to Watson. Next, he considered the current state of phrenology, including 14 objections, to which he found facile answers. For example, to the objection 'That the size of any organ is not the measure of its functional power', he replied that skeptics produced no evidence, while 'We

---

[11] Gibbon 1878, II, 6.
[12] As a model, Stanley published 581 questions, ranging from church history to demography and natural history. Stanley 1836?, edn 2, 1848.
[13] Watson 1836a, v.
[14] Watson 1836a, 120.

have thousands in corroboration'.[15] Then came a town-by-town account of phrenological activities. He wanted to encourage phrenology without inflating its standing. His bibliography included anti-phrenological as well as phrenological writings, though more of the latter. His only real statistics, according to the modern use of the word, were the numbers of phrenologists, societies, lectures, and books. The review of his book in *The Phrenological Journal* was enthusiastic.[16]

The main editor of *The Phrenological Journal* in 1832–36 was a nephew of the Combe brothers, Robert Cox (1810–72).[17] In 1836 he moved to Liverpool and relinquished the editorship. The Combe brothers asked Watson to become the journal's editor and proprietor. Watson was intrigued with the idea, but explained that 'my private income will not allow me to lose much in the Journal, though I shall be content without any profit'. On 11 July 1836 he sent Andrew Combe an outline of proposed changes in the journal's scope:

I should wish annual reports ...
1. On the state and prospects of Education. (Mr. Simpson engaged to supply this.)
2. On the legislative enactments of the year, likely materially to affect the moral state of the community. (Ex: the Poor-law Act.)
3. On advances in our knowledge of the Anatomy, Physiology, and Pathology of the brain: chiefly in a physical view.
4. On the moral and intellectual Statistics of Britain. (Example, Crimes – public subscriptings for moral purposes – temperance societies – library publications.)
5. Progress of Phrenology in Britain, France, America, etc.

He sent still more ideas on 14 September, but by the following 30 March he was still undecided enough to urge George Combe: 'if you can dispose of the *Journal* satisfactorily forthwith, you must do so without any reference to me'. Since that was unlikely, he continued: 'I am to be fixed in Liverpool through May, June and much of July; I shall hope to see a good deal of Mr. Cox; and we may try whether something can be arranged between us ...'. In the next few months Watson agreed to assume both editorship and proprietorship, and he exchanged letters with the Combe brothers concerning the practical aspects of the transfer.[18]

---

[15] Watson 1836a, 73. See also Combe 1827.
[16] Anon. 1836.
[17] On Cox, see Cooter 1984, 279, and Henderson 1887.
[18] Watson to A. Combe, 29 July, 12 Aug., 22 August and 9 September; to G. Combe, 13 October, 18 October, 5 November and 20 November 1837.

He edited five issues of it anonymously, but in his sixth issue he signed his name as editor. Dr Elliotson suggested that *The Phrenological Journal* should publish more case histories, and Watson wrote to George Combe (20 November 1837) that he agreed they were needed. Yet, in his first editorial, he wrote optimistically: 'We safely assume that the general principles of Phrenology are now allowed by the intelligent portion of the public, to be true, useful, and interesting; and that we shall have little further need to defend our subject, in toto, before the public'.[19] He could not maintain such bravado, and in the next issue he was more realistic: 'Ages may perhaps elapse before Phrenology shall be sufficiently advanced towards its maturity, as a science, to entitle it to be ranked on a par ... with ... Natural History and Natural Philosophy.'[20] His analysis of the problems and possible remedies were sensible, but like most advice, it was easier to give than to follow. First, there should be a division of labor – specialization; this was necessary to get beyond superficial aspects of phenomena.[21] Next, there should be a much broader accumulation of facts. He doubted whether the locations of half of the mental organs had been well established, or 'whether the influence of any one organ is at present so exactly known, as to enable us to say how far the actions or character of an individual depend on that particular organ'. Furthermore, who could really distinguish between Self Esteem and Love of Approbation, or between Benevolence and Conscientiousness?

The phrenological investigations he envisioned were probably similar to the botanical investigations he was carrying out: a massive amount of data organized into as many meaningful correlations as possible, in the hope that significant conclusions would somehow emerge. He stressed three major weaknesses in phrenology: standards were often not objective enough, so that too much depended upon the judgment of observers; distinctions were not drawn sharply between facts and their interpretation; and often only favorable facts were collected. Commendable though his zeal and quest for accuracy were, his methodology for phrenology never transcended data collection and correlation.

George and Andrew Combe were pleased with Watson's editing of the first two issues of the *Journal*, which gratified him, but he knew that the caliber of its articles must improve if it were to achieve recognition from the scientific community. Too many authors, he complained to George Combe (3 January 1838), 'expand old ideas instead of compressing them, and have such a turn for long essays, instead of writing short papers with some definite point'. When educator-reformer William B. Hodgson visited him for two days in 1838, Watson found that 'his store of

---

[19] Watson 1837c.
[20] Watson 1838a.
[21] Watson 1838a, 99. The complacency he criticized was one of the faults of phrenology later identified by Lewes 1867, II, 408–9.

information is certainly good', but 'he wants much polish, partly the refinement that springs from the coronal region, partly the conventional refinements of society'. In this case, Watson's low regard for a fellow phrenologist was not widely shared; Hodgson had a very successful academic career.[22] When Watson sought London contacts, it hardly seemed worth the effort. In April he went there to arrange a lecture series for George Combe, but he felt he should report to him (17 April 1838) 'a deplorable account of the jealousies and quackeries of the phrenologists of the Great Metropolis ... They are split into petty coteries, snarling at each other; and the lecturers amongst them, it is said, would regard [G. Combe's] coming with dislike, as an intrusion upon their ground.'

In 1838 Watson exchanged more than a dozen letters with the Combe brothers on publishing *The Phrenological Journal*. He agreed that it contained less general instruction and more information on activities than before he took charge, but he saw no alternative (to G. Combe, 10 June):

> ... since no new discoveries in Phrenology have been made since its Journal came into my hands, there is nothing to tell, except acts done for and against it, and other current events deserving record. Besides this, I was desirous, in a few of the first Nos. of the new Series, to convey an impression that people were busy about Phrenology in various ways and places.

By portraying phrenology in the best possible light, he hoped reality would eventually catch up with his optimistic account.

Watson gave 12 lectures in Manchester sponsored by both the Athenaeum and the Mechanics' Institution. He was a serious lecturer who might hold the attention of some by the cogency of his argument, but none by the elegance of his delivery. He admitted this to George Combe (27 April 1839) while lamenting the shallowness of most listeners: 'only a few surgeons, and others who had studied sciences, saw the bearing of physiology on phrenology, and phrenology on ethics'. The prestigious Royal Manchester Institution defeated an attempt by its Lecture Committee to sponsor additional lectures by Watson,[23] and he continued on to Warrington to give more lectures. He lamented that phrenology made greater progress among the ignorant than among the educated. When George Combe wrote from America complaining that Americans expected lecturers to entertain them,[24] Watson responded (18 November 1839) that the situation was no better in England, where

---

[22] Watson to G. Combe, 18 August 1838. On Hodgson see Cooter 1984, 286, and Gordon 1891.
[23] Wach 1988, 392-3. I owe this reference to Bob Glen.
[24] Combe Papers, vol. 7396, ff. 10-13.

some lecturers sought popularity by depending on lantern images and head-feeling rather than on intellectual rigor.[25] Since he had met with little success, Watson decided against further lectures. Combe regretted this decision, 'because if men of scientific attainments like you decline, the quacks have the whole field'.[26]

Another strategy to boost phrenology was organizing national meetings. When he wrote *Statistics of Phrenology* in 1836, Watson felt uncertain about how to do it,[27] but by 1838 he and a few others decided to organize national meetings in conjunction with annual meetings of the British Association for the Advancement of Science. Some phrenologists were interested in both groups, and phrenologists might also recruit new members from those who came mainly for the science meetings. Watson attended the BAAS meeting at Newcastle upon Tyne in August 1838 and later reported to George Combe (4 September 1838) that he and Sir Walter Trevelyan had taken 'some pains' to collect signatures on a petition to affiliate with BAAS. Watson heard that the Statistical Section had had to exist for two years as a separate association before BAAS reluctantly accepted its affiliation. Watson thought that phrenology would also have to demonstrate that it was 'acceptable to members and the public' before a Phrenological Section could be established. Watson and others tried again the next year at the BAAS meeting in Birmingham (letter to Trevelyan, 26 April 1839). They failed again, and the experience was too painful for Watson to send details to Edinburgh. He attempted to organize a national phrenological association, but responses to his membership circular were discouraging. He told George Combe (19 June 1840) that those who replied 'are uninfluential in the world, and the majority are not likely to be in Glasgow' at the next BAAS meeting. As of 7 July, he still planned to attend the Glasgow meetings himself, but on 12 July he began to prepare George for his absence:

If I do not attend in Glasgow, I will send the needful particulars respecting the state of the Association, and the private business to be done, along with my resignation as Secretary ... The list of members was sent yesterday.

On 23 July he sent advice on when to schedule phrenological meetings to avoid competing with the more popular BAAS, and hinted that Glasgow was too far away and that editing the *Journal* was too time-consuming for him to attend easily. James McClelland invited Watson to lecture in Glasgow and he responded with similar evasions.[28] On 29 July he wrote to George Combe concerning the number of

---

[25] Watson to Combe, 18 November 1839. See also Cooter 1984, 278, 279 and 294.
[26] Combe to Watson, 16 January 1840; quoted in Gibbon 1878, II, 79.
[27] Watson 1836a, 45–6.
[28] Watson to McClelland, 25 July 1840, in Combe Papers.

sessions to plan and what to include in each. A week or so later he received Robert Cox's negative assessment of the prospects for a national phrenological association. Watson replied that, as an advocate of phrenology, he might have deceived himself, but he mostly blamed others, including Cox for his mistaken hopes. Cox was unwilling to accept any blame of his editorship and responded that 'creditable facts were never to the best of my recollection coloured so as to appear more significant than they were in reality'.[29]

An excuse which Watson felt could relieve him from any obligation to attend the Glasgow meetings finally arrived – news of a brother's death. (Zeph had died in India on 10 February; the news reached Hewett six months later.) He sent this excuse, along with further advice on the Glasgow program, to George Combe on 2 September. Some three weeks later Watson confessed (26 September 1840) that he had been alert for any reason not to go. George was annoyed that Watson had defected from a project which he had started. He may have known how little regard Watson had for his brothers, for his reply (28 September 1840) discounted that excuse. Watson responded (11 October 1840) that, because Glasgow phrenologists were active in increasing their numbers, 'I thought the prospect improved; and as the prospect brightened, I felt so much the more at liberty to suit my own convenience.' Even if the numbers of Glasgow phrenologists were increasing, this reply was disingenuous.

In the three volumes of *The Phrenological Journal* which Watson edited (11–13), he expressed his views and standards at great length. All of the book reviews were anonymous and apparently written by him; he wrote the news section and a number of anonymous articles, such as 'Phrenology Supported by Scientific Men'.[30] He had assumed that phrenology was a viable science and that it would only take his comments and firmness to whip it into shape. Even if these assumptions had been correct, his methods were inadequate. He was hypercritical, easily offended, quick to retaliate by reporting to the world the sins of his adversaries, and he wrote no investigative articles that others could use as models. By Watson's third edited volume, his dissatisfaction with the progress of phrenology was blatant. He wrote a tactless tirade on the problems editors face when declining inferior articles, with two examples:

> we have had the misfortune to give grave offence to Mr Levison and Mr Prideaux, both of them active phrenologists (as we hear without personal acquaintance with either), and, likely enough, useful in their respective and proper spheres of action; but neither of them, in our opinion, is likely to have

---

[29] Watson to Cox, 9 August 1840. Combe Papers. Cox's response is quoted from the margin of Watson's letter, but he clearly did send the response to Watson, because Watson's next letter to Cox, 23 September 1840, responds to it.

[30] Watson 1839a.

conferred much benefit on the readers of this *Journal*, or on its credit in the estimation of scientific or philosophical minds, by some of their lucubrations submitted to the editor.[31]

Thomas Prideaux did not accept this drubbing silently. He replied in *Strictures on the Conduct of Hewett Watson, F.L.S., in His Capacity of Editor of the Phrenological Journal*, which showed an understanding of Watson's character (p. 4):

> it is not to be expected that individuals will quietly submit to be visited with the petulances and impertinences of Mr. Watson without retaliating, neither is it desirable that they should do so; indeed, allowing Mr. W. to indulge such a penchant unchecked, would be a great unkindness to him, as it would have a tendency to increase his naturally strong bias towards the belief, that he really is a very redoubtable personage, and thus prove the means of gravely misleading him as to a question of fact. I would hint to Mr. Watson that it is very unwise policy in an individual, whose egregious vanity and conceit render him so peculiarly sensitive, to commence a mode of attack, from which he shrinks when retorted upon himself; and, it is to be hoped that now he has tasted his own physick, and finds its flavour so very unpalatable, he will learn to be a little more sparing in dispensing it to others.

Toward the end of his life, the words of Prideaux and others whom Watson had verbally abused may have haunted him, but in 1840 he had the last (published) word. In reviewing the counter-attack, he explained further Prideaux's fallacious phrenological ideas and asserted the right of editors to publish private letters addressed to them. He thought, 'Mr Prideaux must be callous, indeed, if he cannot yet feel any shame at the unenviable position in which he has compelled us now to place him.' Nevertheless, Watson offered his victim some consolation – the news that he was resigning as editor of the *Journal*.[32]

George Combe and others at various times had warned Watson that his editorials were harsh. He defended them to Combe (27 April 1839) by claiming that his supercilious and caustic tone 'suits many readers, it is echoed by talkers, and so tends to diminish the influence of opponents'. Later he shared other attitudinal thoughts with Combe (18 November 1839): 'anti-phrenologists are to be met with scorn and contempt, and laughed down; whilst, at the same time, efforts are to be made to rescue Phrenology from the embraces of the selfishly cunning, and the good-intentioned zealots'.

---

[31] Watson 1840a. On John L. Levison and Thomas Prideaux, see Cooter 1984, 289 and 293.
[32] Watson 1840b.

On 16 December 1839, Watson asked George to take back *The Phrenological Journal* into his and his Andrew's supervision or to find someone else. He was discouraged because, although there had been a sufficient quantity of articles submitted, few were of high quality, and their subjects were 'too varied and difficult for one person to manage properly'. He thought that if the editorship were divided topically among several experts, the quality of the articles could be raised and circulation increased. He admitted that it would be difficult to find editors 'qualified by knowledge, leisure, and inclination'. Later, he confessed to Combe (14 February 1840) that editorial difficulties were not his only motives: 'My present inclination is to withdraw wholly from active support of Phrenology. I am disappointed – even disgusted with phrenologists; for the more I see of them, the less I like them, always excepting a certain few.'

Obviously, at that point he should also have given up on the national phrenological association, and George Combe should have seen the necessity of his doing so. As Watson moved away from advocacy of phrenology, he came close to attacking it. That he did not do so within the pages of *The Phrenological Journal* while still editor, he attributed to feelings of friendship for George Combe (2 October 1840):

> you are so intimately connected with almost all that is phrenological, that objections against phrenological methods are perhaps often received almost in the light of personal strictures, – for they cannot avoid being in some measure objections against your own acts or ideas.

Toward the end of the last issue he edited, and writing in the third person about himself, Watson explained his reasons for resigning and for having been such a critical editor. From the time of his early interest in phrenology he had had doubts about how well some of its truths were established; suspicion gradually became conviction as he struggled with his editorial duties.[33]

He had attempted to impose high scientific standards upon writings submitted, and if he had succeeded he would have devoted the remainder of his life to phrenology. Since Watson lacked the ability to lead or coerce phrenologists into adopting high scientific standards, he concluded his time was better spent on botany.

His relinquishing *The Phrenological Journal* was a replay, in reverse, of the long negotiations which occurred when he took over. His annoyance over George Combe's vague business dealings resurfaced. George Combe's financial errors were allegedly always in his favor because he was 'strongly endowed with the feeling of

---

[33] Watson 1840d.

Acquisitiveness ... Everybody says so, and your cast bears out the general opinion.' Watson pretended not to understand why Combe reacted negatively to this allegedly objective data. No one was accusing Combe of doing anything 'dishonourable or ungenerous', and Watson had always 'defended you by endeavoring to show that your motives for specific acts had been misconstrued'. The deficiency of Generosity in the Scottish phrenologist was rivaled by the deficiency of Diplomacy in the English one.

They agreed that Robert Cox, who had returned to Edinburgh, should resume the editorship. Watson assured Cox (9 August 1840) that 'circulation will increase more in your hands ... because you will give less offence ...'.

Chapter 5

Outlook and Social Responsibilities, 1835–60

I feel intense desire to raise the low, refine the coarse, defend the oppressed, enlighten the ignorant, and to sweep off all the personal distinctions of society.
                                                    Watson to G. Combe, 22 November 1834

Science and Religion

Watson was not a religious person. In grammar school, he had reacted to being 'drilled with Bibles' with intense hatred.[1] Thus, we might assume that he left the Church of England when he went to Manchester at age 16. Some who left found another church, but he preferred to do without; he found his alternative in phrenology. Phrenology lacked the following and prestige of the Church, but when he joined the movement he believed it 'to be the greatest blessing ever placed within the grasp of man'.[2] George Combe agreed; he had also given up on Christianity, but he remained a deist. For Watson, deism was merely a hindrance to progress.

When Combe sent him a new edition of his *Constitution of Man* in 1835, Watson thanked him but regretted that Combe had stopped short of being as effective as he might have been (2 September 1835):

> The grand stumbling block is Religion. Here we altogether differ. You assume an identity [of God as cause of intelligence] & fancy nature affords proof of such. I do not see this proof; do not see the necessity for such an Existence ... you are at times led into false conclusions, arising out of this very assumption. Supposing the book utterly without admixture of religion, natural or otherwise, it would be ... noble.

In the following year, when Watson wrote *Statistics of Phrenology*, he wanted to refer readers to a philosophy of phrenology but decided he could not endorse *The Constitution of Man*. He explained why not to Combe (23 July 1836):

---

[1] Watson 1839c, 266.
[2] Watson quoting himself to G. Combe, 1 March 1830.

we do not discover that any creature has received a constitution; neither does the existence of objects prove a maker, nor their manner of existence *prove* wisdom and benevolence in that maker, if he does exist. First, we have no proof of his existence; secondly, we have no proof that his qualities, if he does exist, are the same as those of human beings; and thirdly, the constitution and relations of matter appear to exist as inevitably as matter itself, and to be inseparable from it. So far as ascertained science goes, every thing countenances the presumption that matter, its constitution and its relations, never commenced and will never end.

When Watson wrote a reply to William Scott's attack on Combe's *Constitution of Man*, he complained to Combe (26 November 1836) that 'a *complete* reply is impossible without offending religious opinions'. He seems to have known that he could share these opinions with Combe without losing his friendship. In 1839 he abandoned caution and admitted in a natural history journal that he was 'a renegade from Toryism, and a Dissenter from the Church of England'.[3]

In early 1843 Combe was working on the fifth edition of his *System of Phrenology* and sent revisions to Watson, who cautioned (28 February 1843) that Combe could not establish arbitrary rules of evidence for phrenology and then call it a science: 'It is directly at variance with Herschell's definition of science.'[4] Combe sent along another batch of revisions and Watson responded (7 March 1843) with the comment that one needs to distinguish between 'Those who have merely announced truths' and 'those who have proved such truths ... by a series of careful observations and well conceived experiments'. Combe sent thanks (10 March 1843) but happily reported that he had obtained more supportive comments from Dr William Gregory, Professor of Chemistry in Aberdeen. Combe decided that 'Such wide differences of opinion as exist between you & Professor Gregory, both men of science & both lovers of truth, can be explained only by the differences in your brains & habits ...'.

Watson held up high scientific standards for Combe in the hope of still salvaging a respectable phrenological science – some day. Perversely, Watson asserted (23 December 1843) to a fellow botanist, Charles Babington, that phrenology was a more important subject than botany.

Combe published in *The Phrenological Journal* some extracts from his forthcoming *Remarks on National Education* (1847), and asked for Watson's opinion. Combe's deistic language prompted Watson to imagine a religion based on verifiable knowledge (23 January 1847):

---

[3] Watson 1839c, 269.
[4] John Herschel (not Herschell) 1830.

[it] will discard all that is supernatural, and rest only on what is natural, that is, on knowledge and experience. Such being the case, the Expounders and teachers of religion (for present life) will be the men of science – those who investigate nature. And this class of men is notoriously disinclined to venerate and worship a Deity.

Watson's claim that men of science were disinclined toward religion seems impossible to document. Those who felt that way were likely to be reluctant to admit it in print, and those who did discuss science and religion were generally concerned to minimize the conflicts in order to minimize opposition to science.[5] Watson decided that the naturalistic approach would be fatal to religion: 'instead of knowledge and religion entering into partnership, the former will gradually smother and destroy the latter'. Combe's convictions were as firm as Watson's, but he complemented Watson (28 January 1847) for remarks that were 'calm, clear, and to the point'. Since they were able to discuss religion without giving offense, they continued to do so for the rest of that year and the early part of the next, but without either one changing his outlook. Their reason for persisting was that Combe was writing a pamphlet, *On the Relation between Religion and Science*.[6] After reading it, Watson indulged in a rare bit of levity (14 May 1847):

One is tempted to think you are to be the prophet, founder, and promulgator of a new religion, Combe-ism, or to place religion among or upon the sciences. When the present Bishop of Norwich [Edward Stanley] was raised to the clerical bench, the Tory papers said he was learned in all the ologies except The-ology, which he ought to have. You would make learning in all the ologies into something very like The-ology of the ologies – the religion of science, or science of religion.

Watson's evolutionary beliefs influenced his cosmology, as is evident in one response to Combe's inquiries (18 February 1848):

You say that you cannot see a process of reasoning by which I could go back from the conclusion once held 'that design indicates a Designer'. The difficulty is in the word used here. Earlier I fell into the false logic of assuming that adaptation is design. Design implies a Designer; – but Adaptation does not necessarily imply a

---

[5] Brooke 1991. Budd 1977. Eisen & Lightman 1984. Gillispie 1951. Lindberg & Numbers 1986, chaps 12–14. Mullen 1988. Royle 1974. Ruse 1975; Ruse 1979, chap. 3. Turner 1993.
[6] 45 pp. in 1st edn, but expanded until it reached book length in 4th edn (1857). Gibbon 1878, II, 354–7. Extracts from Combe 1847b (1–8 and 12–13) are reprinted in Moore 1988a, 409–14.

Designer. It is Adaptation, not Design, which is the fact recognized by me in nature. The word design expresses the simple fact, together with a causal explanation of it, which is assumed. The word adaptation expresses the simple fact only, and assumes nothing but the reality of our own mental ideas of external nature.

As he became more and more critical of phrenology's scientific pretensions, Watson compared it unfavorably to other sciences. Although phrenology seemed more interesting than astronomy and geology (he wrote to Combe, 30 April 1854), these sciences made more rapid progress: 'Is it not remarkable that Phrenology should have so far died away, while the tendency to scientific studies is undoubtedly on the increase?' (3 February 1857) And, more devastatingly (20 February 1857):

> Geology and Phrenology, in their modern conditions, are sciences of recent date, and not very unequal in age. Both were vehemently opposed at first as anti-scriptural, untrue, &c., &c. Both are sciences of reasoning upon facts. Here the parallel stops ... In geology the opposition has gradually changed into acquiescence. Geology is now perhaps at once the most popular and the most philosophical of the sciences. It has immensely advanced in the past quarter of a century, both by progress and by diffusion. Accumulation of new facts in almost countless numbers, and re-examination and reconsideration of the numerous older facts, have placed Geology in this advantageous position. On the contrary, in Phrenology opposition has changed into neglect or contempt. Phrenologists are not recognized, as such, by their fellows of science. Phrenology has made no progress. There is no accumulation of new facts, and no reconsideration of old facts.

Nevertheless, he assured Combe that 'Phrenology has been of great benefit to yourself, and it has benefitted the public thro' you ...'.

Sometimes they turned from cosmological to practical aspects of religion. Both were interested in educational reform, and this subject led to comments on religion, because religion was a part of almost all British education. In 1857 Combe sent Watson his latest pamphlet, *On Teaching Physiology and Its Applications in Common Schools*. Watson complained as usual about its theology (9 November 1857), but nevertheless decided to send copies to friends and relatives.

Political and Social Views

For most people swept up in the phrenology movement, expectation of social reform was part of the appeal. Since two of Watson's most important intellectual commitments – first phrenology, and by 1834 evolution – were somewhat radical,[7] one might expect that some radicalism carried over into his political and social outlook. He made only a few private pronouncements that sound radical (for example, 'I feel intense desire to raise the low ... and to sweep off all the personal distinctions of society ...', to Combe, 22 November 1834). However, his sincerity in such pronouncements is suspect, for he lived on inherited wealth. He was generous with money, but not to the point of becoming poor enough to have to work for a wage. He commented on political issues of the day in letters to Combe. The Reform Bill of 1832 did not achieve as much as reformers wanted, and some historians now see its main significance as being the beginning of a decade of reform rather than being a great achievement in itself.[8] Its passage coincided with George Combe's efforts to assist a sculptor find work. Watson was willing to contribute a substantial amount to aid Lawrence Macdonald's work, but he disliked killing two birds with one stone: 'I would subscribe to celebrate the Reform Bill by founding schools, but not by founding monuments, altho' the latter does give employment.'[9]

Even in a decade of reform, Watson was somewhat pessimistic about what could be achieved. He liked an article on penitentiaries and treatment of criminals in *The Phrenological Journal* which he presumed George Combe had written, and he complemented him on it (29 August 1834). However, Watson felt that the enlightened methods advocated would never be tried during their lifetime. That being the case, 'is it not fitting to extirpate the worst for the benefit of the community?' He admitted that this was a disagreeable alternative but thought it 'better than tying up wolves and tigers for a time, & then turning them out on society with fresh appetite for blood & destruction'. After expressing these unliberal thoughts, he sheepishly acknowledged that Combe might smile in derision.

His pessimism was not dissipated by Combe's soothing response, but it did inspire him (22 November 1834) to contrast what he would like to do with what he actually did – 'I now shun society, disgusted with the selfishness, the rivalry, the petty jealousies & animosities, the duplicity & knavery ...'. However, his depressed feelings were occasional rather than chronic. During the 1830s, while he pinned his main hopes on phrenology as the salvation of society, Sir George Mackenzie persuaded him to send a political-phrenological testimonial to Lord Glenelg, Secretary for the Colonies (18 March 1836):

---

[7] On the relationship between evolution and radicalism, see A. Desmond 1987 and 1989.
[8] Gash 1979, chap. 5; Llewellyn 1972; Woodbridge 1970; Woodward 1962, books 1 and 4. Bob Glen thinks this conclusion remains debatable.
[9] Watson to Combe, 30 June 1832. Watson was responding to what may only have been a hypothetical idea, not one which Combe was urging. Watson discussed his financial contributions to Macdonald in letters to Combe on 27 August 1832 and 29 August 1834. On both the Reform Bill and Macdonald, see Gibbon 1878, I, 249–52.

being convinced, after several years of careful attention to the subject, that it is quite possible to determine the disposition of men, by an inspection of their heads, with so much precision as to render a knowledge of Phrenology of the utmost importance to persons, whose duties involve the care and management of Criminals.[10]

Mackenzie's idea which Watson supported seems to have been that phrenology could be used to help decide which convicts to send to Australia.

While editor of *The Phrenological Journal*, Watson had wanted to include annual reports on progress in education, legislation and criminal statistics, and he wanted them discussed without involving party politics. However, his ability to implement these plans was quite limited.

One of the leading Scottish phrenologists was James Simpson (d. 1853, aged 73), an advocate (lawyer), who worked closely with the Combe brothers. In 1839 Simpson came to London and gave a series of lectures. Simpson wanted the government to adopt some of his educational ideas, and to build support for them he ignored his own phrenological ties. This displeased Watson, who went to hear him and afterwards accused him of 'appropriating the ideas of phrenological writers, without due acknowledgment'. Watson characterized Simpson's response to George Combe (24 June 1839) as riding a high horse:

He couldn't help it if phrenologists did object. He was doing more good by not speaking out. His lectures were on Education, not Phrenology, &c. &c. Dr A. Combe writes me that he has had some conversation with Mr Simpson since his return to Edinburgh; and with the advantage of the government scheme falling to the ground in the interim, between his own and my remarks to Mr S. The effect, I think, will be to make Mr S. feel that he cannot lose the good-will of phrenologists; for in losing them, he will lose his audiences to at least the half; and in losing his audience he will lose the compliments and countenance of the M.P.s, &c.

Watson's frustration with the slowness of British reform did not cause him to idealize America's political system. Indeed, his views on America were probably no more favorable than those of his conservative sisters. He sent Combe (18 November 1839) in Maine his sympathy for having suffered the discomforts of American Hotels, and he confessed to having 'a sort of antipathy to anything American, and

---

[10] From a copy in the G. Combe correspondence. Lord Glenelg = Charles Grant, Colonial Secretary, April 1835–February 1839; Rapson 1890. Sir George Steuart Mackenzie (1780–1848) was a mineralogist and one of the most prominent phrenologists; de Giustino 1975, 49–51 *et passim*; Gibbon 1878, I, 109, 135 *et passim*; Hewins 1893.

cannot reason myself out of a feeling akin to disgust, albeit that is rather too strong a term to apply, towards Americans'.

In the same letter he announced: 'I have resolved not again to lecture on Phrenology; – partly because it is painful to my own self-estimation, to feel conscious of shying the truth in the theological bearings; and partly because lecturing is not agreeable to my taste, except under the consciousness of producing some actual good ...'. And in a similar vein, he admitted:

> The remark of Mr. Allen, about the new series of *The Phrenological Journal* being 'less moral' than the former, is in a certain sense correct. There is more rudeness now, and less consideration for the effect to be produced on the feelings of individuals, by remarks made. But let me qualify this admission by adding, that the difference is at least in part attributable to the smaller influence of Secretiveness and Love of Approbation in my disposition; and that from earliest childhood I have never been able to act a character.

His self-analysis reveals a desire to understand himself, within a phrenological context, and he failed either to realize that the phrenological context might be inadequate for the task or to reckon with the possibility of his mind evading self-analysis.

The problem of sincerity v. hypocrisy is relevant to his political outlook because in practise he was less committed to 'sweeping off all the personal distinctions of society' than he believed. His ghost-written biographical sketch (1839) proudly announced at the end that he was a renegade from Toryism, and a Dissenter from the Church of England, yet this is not the impression made by the first two pages of text (out of six), which are devoted to his illustrious ancestry. Those two pages itemize aristocrats and churchmen, whose blood mixed to produce the unique Hewett Cottrell Watson.[11] Years later, in the context of personal qualities rather than politics, he admitted (2 February 1847) to Combe: 'I do attach some importance to birth, and a little value to descent.'

Perhaps it was his eye for ancestry that led Watson to withhold his approval of Richard Cobden, who otherwise seemed the sort of politician who would win his support. From humble origins on a farm he rose to become a successful textile manufacturer in Manchester. In 1838 he had been a founder and the most important member of the Anti-Corn-Law League, and he was elected to Parliament in 1841, where he was both popular and effective.[12] Furthermore, he was receptive to Combe's teachings on both phrenology and education. When Combe asked for

---

[11] Watson 1839c.
[12] On Cobden (1804–65), see Edsall 1986, Hinde 1987 and Morley 1887. For a brief sketch of the Anti-Corn-Law League, see Spall 1988.

Watson's comments on the seventh edition of *The Constitution of Man*, Watson responded (27 December 1847) with a miscellany of reactions, including this one:

> You mis-praise Mr Cobden. Doubtless he is a superior man; but neither his doings nor his speeches will bear the rigid scrutiny for truth and conscientiousness.

Combe was quick to defend an admirer (16 January 1848). Watson reiterated (23 January 1848) various suspicions, but also beat a tentative retreat: 'though Cobden is an astute man, and not too sincere, he is a most useful and valuable personage ...'. Combe (11 February 1848) was glad to see these concessions, but Watson was not finished. He admitted (18 February 1848) that he had never seen Cobden, but his speeches show 'rather too much a partisan and case-maker to be held a strictly conscientious man'. Watson feared that 'all successful public men swerve a little from truth and reality, and would be unsuccessful if they did not'.

Watson 'was delighted with the expulsion of Louis Philippe' (written to Combe, 14 March 1848), but wondered how much better off France would be without him. Combe asked two days later how much better the situation was in Britain. Watson answered with a sweeping plan of 12 steps for organizing the middle classes into a National Association of political committees which would save the country from the misguidance of the other classes by constantly monitoring public opinion and urging government to act in accord with it. The whole endeavor should be run by 'practical men such as Mr Cobden'. In two-and-a-half months Watson had progressed from warning about the unreliable Cobden to entrusting the government to him!

Combe (23 March 1848) was thrilled by Watson's elaborate plan, for it seemed compatible with Lord Dunfermline's scheme of an Association for Scotland to promote non-sectarian education.[13] Combe's initial impression that Watson and Dunfermline were on the same track was corrected by Robert Chambers and other friends, who persuaded him that Watson was talking politics, whereas Dunfermline's project was non-political. (Chambers was willing to become involved provided they started by organizing labor.[14])

Combe's excitement and the political climate stimulated Watson to turn his long political letter into a pamphlet, *Public Opinion; or Safe Revolution, Through Self Representation* (May 1848, 24 pp.). Ending government mismanagement had been the theme of his letter; averting revolution became the theme of his pamphlet. Aristocrats should welcome his plan, because merely by surrendering power to the middle class, the latter could save them from 'a bloody collision with the

---

[13] Lord Dunfermline = James Abercromby (1776–1858) was speaker of the House of Commons, 1835–39; Henderson 1885.

[14] Gibbon 1878, II, 243–4.

uneducated masses'.[15] (The liberal Sir Walter Calverley Trevelyan was not alarmed by the plan, judging from Watson's letter to him, 13 May 1848.) It is less clear why 'the uneducated masses' would be content to have government run by the middle class. Aristocrats were often largely responsible for laws protecting factory workers from middle-class bosses. Watson's own upper-middle-class father had prospered by helping business and professional men oppress workers. Watson imagined that his National Association would cause political parties to disappear, and that this would benefit the country.[16] He told Combe (8 May 1848) that he had sent copies of his pamphlet to a half-dozen friends and would only publish it if a majority approved. They all approved, provided he made changes (to Combe, 11 May 1848), but their suggestions were contradictory, which allowed him to choose among them. He appreciated (20 May 1848) Combe's offer to distribute copies of *Public Opinion* to Scottish newspapers, yet Watson did not undertake any massive distribution himself. He admitted to Combe (15 June 1848) that he had sent out only a few copies to friends and to Peel, Hume and Cobden, and that he left it up to his publisher to send copies to editors of the press. He was competing with other reformers, and his pamphlet did not take the country by storm. He was therefore willing to go along with whatever reforms won out. When Cobden's ally, Joseph Hume, developed a scheme which seemed more likely to succeed than his own, Watson was willing to go 'from house to house of the trades people and gentry, to request their signatures' to support it.[17] He mentioned this endeavor to Combe because 'you would hardly have suspected me, as one likely to undertake so troublesome and personally unpleasant a task'.

Combe replied (24 June 1848) that he did get Watson's pamphlet reviewed in *The Scotsman* (24 June 1848), and he was glad that Watson collected signatures on a petition for Hume. Combe had invited to dinner Alexander Russel, editor of *The Scotsman*, Robert Chambers, a publisher, and Richard Shaen, a Unitarian minister. Combe and Russel defended middle-class control, while Chambers and Shaen urged labor's participation. Watson thanked (26 June 1848) Combe for the review, and agreed with Combe and Russel that greater participation by the laboring class in the management of the economy and government must await its being better educated. But Watson 'despair[ed] of any good system of national education while we have a state-church'.

Watson's best political pronouncements were not made when he wrote sweeping solutions to government crises, but when he responded to someone else's notions.

---

[15] Watson 1848a, 3.
[16] Watson 1848a, 23.
[17] Watson to Combe, 15 June 1848. On Hume, see Hamilton 1891.

Trevelyan believed that alcohol was the cause of many social ills, and that its sale and consumption should be banned. Watson disagreed (26 November ?):

> We cannot expect in our time to see a majority of the nation in support of such a law. And it would surely be a political inexpediency for a minority to enact & execute a coercive measure against a large majority in a matter where there would be comparatively few neutrals or indifferents. If the time should ever come, when a decided majority is in favor of such a law, just in like proportion will the need for it have lessened.

George and Cecilia Combe visited Dublin in 1848, where they met Lord John and Lady Frances Russell. Lady Frances asked George Combe to examine phrenologically the heads of her stepdaughter and son – the IQ test of the day. He obliged, but was more interested in her husband, whose head he described in his journal and in a letter to Watson.[18] Russell's bust had been described already in *The Phrenological Journal* (vol. 12, p. 268), and Watson thought (29 September 1849) that Combe's assessment of him (rather mediocre) accorded with the description of the bust. Watson also commented that Lord John's Potato Commission to Ireland 'is said to have cost the country £54,000; while it would be difficult to prove 54s worth of benefit from it, – unless to Lord John who required a certain kind of report, for his own views, which of course they made for him'. Although Watson's estimate of the value of the Potato Commission's report agrees with that of later historians, the commission had been appointed by Robert Peel in 1845 and had submitted three reports to him in that year. Russell only came into office in July 1846.[19]

Combe wrote out a plan for increasing British self-rule and obtained Watson's reaction (14 January 1849):

> Local legislation for local objects appears a desideratum to a great extent now, not only because Britons are becoming more competent for self-rule, but because the practice of self-rule must tend strongly to increase the competency. With your essential principle of legislating for Ireland, in Dublin, in Scotland, in Edinburgh, for England, in London; – in all matters not imperial, I can cordially concur ...
>
> But, [my] present impression is adverse to the mode of tri-viding the Commons, suggested by you.

---

[18] Combe's journal entry for 7 September 1848 is quoted by Gibbon 1878, II, 270–1. Combe's letter to Watson is dated 27 September 1848. On Russell, see Fraser 1988 and Rae 1897.

[19] Large 1940, chap. 1; Munsell 1988; Woodham-Smith 1962, 44–7 and 424.

On a related theme, he added that Cobden might 'be borne into the Premier's Seat' if 'he would take a decided position, as leader of the liberals'. A few months later, Watson decided approvingly (18 April 1849) that: 'Cobden has now fairly taken the position from which he cannot draw back without discredit, and I am glad to see him there.'

Combe (28 October 1853) notified Watson that Cobden had plans for establishing a secular school:

> like the Birkbeck schools, at Manchester, excluding religious creeds, and will ask the Government to give it the same support they afford to sectarian schools, or to have a reason why they decline. He knows that they will decline; & then he will start, in Parliament & out of it, an agitation that shall never rest till a victory is won, or the Nation shall abandon it outright. Mr William Ellis, in June last, offered him £5000 to institute the school, if he would take this line, & he, Mr Cobden, writes me that he has obtained already in Manchester subscriptions to the amount of £500 a year, for three years certain, to maintain it.

Since one of Watson's constant laments was the lack of secular schools, Combe expected a jubilant response.[20] Watson had also wanted Cobden to become prime minister – a strong recommendation for a school's sponsor. But his approvals had last been expressed in 1849. Now (30 October 1853) Watson reverted to his earlier suspicions:

> You and I estimate Mr Cobden quite differently in one point of character. I regard him as essentially unscrupulous; that is, with a feeble feeling of Conscientiousness; but, at the same time, too clever and shrewd, too circumspect and fore-sighted, to let this natural deficiency become very obvious. Yet some persons who note his public doings, detect the deficiency, as I know by inquiry among men who are competent to judge, and have had the opportunities, both in London and Manchester. Nevertheless, he is a most useful man for the age, which, as you know, is not very anxious for the rigidly right.

Britain began drifting toward war with Russia, and the tireless Combe took up pen to oppose war. He sent his *Apology for the Peace Congress*[21] to Watson, who responded (14 December 1853) that it:

---

[20] Combe's importance in education is seen in the gigantic reprint of his writings on education in 1879. On the slow pace of British reform of schools, see Novo 1988. Around 1855, the British Association for the Advancement of Science began to press for more science in the schools; Layton 1981.

[21] The sixth Peace Congress since the revolutions of 1848 met in Edinburgh, 12–16 October 1853; Hinde 1987, 245–6 and 345; Edsall 1986, 262–3. Cobden was a major speaker at the Edinburgh congress.

Figure 4 George Combe (1857), by John Watson Gordon.

reads to me sound as far as you go; and pitting England against America, or France. But how [to] appeal to the people of Russia! And even with its educated Emperor, the early drawn swords of France and England would have kept the peace towards Turkey far more surely than moral appeals. Pity, that many of the Peace-men do not distinguish between what is morally right and what is presently possible.

About three months later (9 March 1854), as Britain moved closer to an alliance with France and Turkey, Watson wrote to Combe that he wished he could 'attribute the hostility of the Cabinet towards Russia, to higher motives than self-interest and offended pride, – the fear of losing office and desire to punish the man who has over-reached them'. Yet he was no more satisfied by those who opposed the Crimean War, for he complained (26 December 1854) to Combe that 'the Peace men seem to me to be rather degrading their own principles by the mode in which they argue for or urge them upon others'. This broad condemnation could not fairly include Cobden, whose speech in the House of Commons on 22 December was his first on the subject since Britain had begun to fight, and he merely argued that Britain could achieve its objectives through diplomacy.[22]

The ever-busy Combe sent Watson his Remarks on the *Principles of Criminal Legislation and the Practice of Prison Discipline* when it appeared, and the latter responded to some of Combe's ideas using himself as an example (25 May 1854):

... the Laws as they are, by enactment, example, &c., actually prevent a vast deal of crime. Say that upwards of 7,000 persons were committed in 5 years (page 29) for certain crimes, that is, were not deterred from crime. How do you know that 70,000 others, or even 700,000 others were not deterred by the Laws? I am naturally very destructive. What prevents me from killing those who injure or offend me? The legal consequences, more or less aided by the social conventions. I really believe this of myself, altho' benevolence is certainly a vivid and a rapidly-roused feeling with me.

I am not sure that your refusal to punish for example sake, will fully harmonize with views about the use of pain and suffering, in nature, put forth in [your] *Constitution of Man*?

Combe had developed an early interest in finance, and he concluded that over-reliance on paper money was dangerous for the economy. In 1856 he published a

---

[22] On Cobden's speech, see Edsall 1986, 274–5. Combe was probably more deeply concerned than Watson; see his letter on it to Cobden, 6 January 1855, quoted by Gibbon 1878, II, 336–9.

pamphlet on the subject.[23] Watson was pleased to have this readable treatise to which he could refer people. His comments (12 April 1856) show that in becoming a man of science he had not abandoned worldly interests. Combe was able to utilize Watson's advice because the pamphlet went through ten editions (under a new title).[24]

A mutiny of native troops in the British Indian Army began in May 1857, and was not fully suppressed until February 1859.[25] Watson had two brothers who had died in that army, and in 1834 his sister Frances Harriett had accompanied her husband, Captain James Spence, to India for his service in it that lasted eight years or more.[26] However, family involvements did not incline Watson to view the British record in India with indulgence. He remarked to Combe on 9 November 1857:

> What a retribution in India for the long course of injustice, injury, and insult inflicted on the natives by Englishmen! But when we thus see the retributive suffering miss the most guilty altogether, and fall heaviest on the innocent or slightly implicated – it does seem to me impossible to look upon the requital as a justice, though it is a moral consequence of fault. We execrate the injured Nana [Sahib], for striking at the nearest representative of the Indian Government he can reach. Yet in our theological philosophies, we feign this infliction of vicarious suffering to be natural justice, providential justice.

Combe agreed; he had written against British oppression of India in the seventh edition of *The Constitution of Man* (1847) and would write further on the subject in the *Scotsman* (February 1858).[27]

Pacificism to Watson did not mean being militarily unprepared. When Trevelyan complained about volunteer militia to him, Watson offered this perspective (1 August 1860):

> Though equally with yourself regarding war-making as sheer folly ... I still look with some satisfaction on the Rifle mania, & believe Lord Palmerston pretty correct in designating the Riflemen efficient members of a Peace-Society. I look upon the present Emperor of the French [Napoleon III] as being so thoroughly unscrupulous & ambitious that he would delight in invading this country on any

---

[23] *Refutations Refuted: A Reply to Pamphlets put forward in answer to 'The Currency Question Considered'*; Gibbon 1878, II, 272–4.

[24] Gibbon 1878, II, 393.

[25] Hibbert 1978; Zastoupil 1988.

[26] On his brothers, see above, chaps 1 and 4. In a letter to Elizabeth Dawson, 19 June 1834, Julia Watson Gorton stated that Frances Harriett was going to India for six or seven years, but Watson's letter to W. Hooker, 24 March 1842, indicates she was still there.

[27] Gibbon 1878, II, 359–60.

pretense, if he could do so successfully. Trained riflemen will enter rather largely into his reasons for abstention.

One wonders whether Watson was conscious in expressing these sentiments of inching closer to his father's perspective. A remaining difference was that Hewett presumably would not have sanctioned the use of militia to prevent workers from striking and organizing.

When Watson wrote in 1834 that he wanted to 'sweep off all the personal distinctions of society', he was thinking of undermining aristocratic privilege, not of advancing women's liberation. He disapproved of his sisters because their involvement with housekeeping was less than his mother's, and that attitude never changed. He remarked in his last known letter to Combe (27 November 1857): 'There is a feature in the lists of Shareholders in J.S. Banks that I do not like, the increasing number of Lady Shareholders and "Executors of".'

More commendable from a modern viewpoint was Watson's generosity on many occasions.[28] When he and his sisters left Congleton, they agreed to send money to the widow Elizabeth Dawson for the rest of her life. Because of the smallness of the sums and the fact that Mrs. Dawson and the witnesses to her will were all unable to sign their names, one assumes that she was their housekeeper when they were children. Providing a small pension to one's loyal housekeeper may not be an outstanding example of generosity, but at least Hewett and his sisters conscientiously sent her their sums regularly during eleven years or more.[29]

---

[28] See, for example, letters to Combe, 7 October 1853 & 26 April 1856. Watson's epitaph says: 'It is better to give than to receive.'

[29] The Elizabeth Dawson collection at Congleton consists of her will, dated 11 July 1836, and letters, mostly from the Watson siblings. The only one from Hewett is dated 26 January 1836.

Chapter 6

Continuing Plant Geography Studies, 1833–48

> ... Watson was as near the truth as anyone has been or is likely to be when he observed as early as 1835 that 'on the average a single county appears to contain nearly one half the total number of species in Britain' ...
>
> Dony 1963, 384

In Britain

For a few years after he moved to Thames Ditton, Watson was even more active in his botanical studies than he was in phrenology, but during 1836 he began to reverse these priorities, and after the second volume of his *New Botanist's Guide* appeared in 1837, he published little on botany until 1841.

In 1833 he compensated for his neglect of soils during the previous summer by examining the importance of rock substrate for determining the distribution of species. (His reasoning in this short essay was so astute that in 1954 an English ecologist reprinted Watson's conclusions in *Ecology* to remind his colleagues of a forgotten pioneer.[1]) Watson detected weaknesses in the arguments of earlier investigators on both sides of the question of whether soils are important for explaining species distributions. In attempting to settle the matter, however, he did not appeal to experimentation. Rather, he relied upon his own experience and judgment and the conclusions of other plant geographers. His unfortunate omission of evidence was because it 'would be far too voluminous for insertion' in J.C. Loudon's *Magazine of Natural History*. There were, of course, scientific journals like the *Philosophical Transactions of the Royal Society of London* or Jameson's *Edinburgh New Philosophical Journal* that published long articles filled with data. However, the papers to which Watson referred were in Loudon's *Magazine*, and he chose to respond in the same journal. Its contributors and readers were mostly amateur naturalists, as he himself commented to Hooker,[2] and its articles were short narratives or descriptions.

---

[1] Gorham 1954; I am indebted to Robert P. McIntosh for this reference. Watson 1833b.
[2] Watson to Hooker, 6 January 1836.

In this case, Watson allowed his scientific standards to be set by amateurs rather than by Humboldt. (Although in Britain some amateurs did professional-quality work while some professionals performed at an amateurish level, it still seems true that most amateurs were more superficial and had less expertise than most professionals.) There are three reasons why he might have done so. First, he showed limited respect for the previous discussants of the question; they must not have seemed worthy of elaborate refutation. Second, he was still a relatively inexperienced naturalist who (despite Humboldt) may not have appreciated the need for thorough documentation. And third, the standards he accepted for this paper were about the same as those followed in his other science, phrenology. There are parallels between this paper on plants and some which he wrote on phrenology. This is evident in his conclusions, which discuss the relative importance, under different conditions, of five environmental factors:

1. Temperature; 2. Moisture; 3. Configuration of the surface (chiefly with relation to shelter and exposure); 4. The mechanical and chemical properties of the surface soil; 5. The mechanical and chemical properties of the subjacent rocks.[3]

Although his evaluation of the interactions of these factors seems astute, his own investigations only indicated reasonable possibilities. Even if one accepted his conclusions, they would be difficult for others to apply without detailed instructions. He did include his new data in *Remarks on the Geographical Distribution of British Plants; chiefly in connection with Latitude, Elevation, and Climate* (1835), which was a revised and enlarged version of his *Outlines* (1832). *Remarks* was favorably reviewed[4] and received the compliment of a German translation in 1837.

Watson was always fond of pointing out the mistakes of other botanists, and in *Remarks* he emphasized their disagreement over the total number of British species – ranging from Samuel Frederick Gray's high estimate of 1636 (published 1821) to George A.W. Arnott's low estimate of 1503 (published 1834). Watson's own estimate was 1470 species.[5] His discussion of this point has greater significance than merely deciding who was right. In 1834 he had written to Winch that he had been won over to a belief in the transmutation of species.[6] The discussion in *Remarks* (1835) was

---

[3] Watson 1833b, 426–7.
[4] The review is undoubtedly by W. Hooker (1835).
[5] Watson 1835a, 39–40. On George A.W. Arnott (1799–1868), see Desmond 1977, 19, and Stafleu & Cowan 1976–92, I, 67–9. On Samuel Frederick Gray (1766–1828), see Desmond 1977, 265.
[6] Watson to Winch, 7 October 1834; quoted below, chap. 12.

his first published inquiry into the stability of species. Admittedly, his approach to the subject was indirect, but since Lyell had in 1832 discredited Lamarck's defense of transmutation,[7] Watson must have felt that there would be little gained from an open declaration until he had either new evidence or a better occasion.

Watson also anticipated in *Remarks* some modern ideas on the statistics of species–area relationships. This discussion did not speak to contemporary concerns to the same extent that his comments on the status of species did. However, the fact that he anticipated our modern interest in such relationships shows that his assessments of phytogeographical issues was often very good. He judged that:

> On the average, a single county appears to contain nearly one half the whole number of species found in Britain; and it would, perhaps, not be a very erroneous guess to say that a single mile may contain half the species of a county.[8]

Joseph Hooker reminded Darwin of this conclusion in 1846, and a modern student of plant diversity, J.G. Dony, thinks Watson was 'as near the truth as anyone has been or is likely to be, when he' made the above statement. Dony pointed out that Watson's verbal observation of the relationship between diversity and size of area could, 'by the use of a logarithmic scale ... be made a linear one'.[9]

Watson's publications established him as the leading British student of plant geography.[10] He was the obvious botanical reviewer of Hugh Murray's *Encyclopaedia of Geography* (1834), which contained William Hooker's discussion of plant geography. Hooker may have written his chapter before Watson published his *Outlines* (1832), since it made no use of Watson's work. It began with a general discussion that was little more than an expanded translation of de Candolle's 'Géographie botanique' (1820), and Hooker based his discussion of British plant distributions on studies by Nathaniel Winch and James Farquharson. Watson's review, published in one of Hooker's journals,[11] opened and closed with complimentary remarks, but the body of his review corrected the data from Winch

---

[7] Lyell 1830–33, II, chaps 1–12. Coleman 1962.

[8] Watson 1835a, 41–2; cf. 1848s. Joseph Hooker called Watson's observations on this point to Darwin's attention, 28 September 1846; Darwin 1985–, III, 342.

[9] Hooker to Darwin, 28 September 1846; quoted in Darwin 1985–, III, 342. Dony 1963, 384. I owe three debts for references here: Charles C. Mann called my attention to a citation to Watson 1835a in Connor & McCoy 1979, pp. 793–4; Earl D. McCoy then provided reference to Dony's article, and David E. Allen, reading this note, informed me that he had brought Watson's writings to Dony's attention in the first place.

[10] Egerton 1979. See also Stafleu & Cowan 1976–92, VII, 98–101.

[11] Watson 1835d; see also Watson to Hooker 27 November 1835. On Rev. James Farquharson (1781–1843), see Desmond 1977, 217.

and Farquharson. Watson's review contained an implicit message that amateur naturalists who reported on local floras needed to be more precise. To assist them, he published two papers: 'On the Construction of Maps for Illustrating the Distribution of Plants' (1836) and 'On the Construction of a Local Flora' (1837). Simultaneously, he published *The New Botanist's Guide to the Localities of the Rarer Plants of Britain* (2 vols, 1835–37), based upon Turner and Dillwyn's, explaining in the introduction that 'the present Guide contains only details of stations [habitats] of more immediate use to collectors'. By 'collectors' he might have meant hobbyists, though he compiled his guide to refine the accuracy of the data for science. By this time, however, he was abandoning botany in order to devote his full time to phrenology.[12]

In the Azores Islands

When Watson wrote his prize essay on plant geography at Edinburgh in 1831, the scope had encompassed much of the Old World, but when he began publishing on botany he limited himself to Britain. Nevertheless, in trading plant specimens he accepted some from various places beyond Britain. He also retained an interest in the relationship between British and foreign species. The latter interest alone might not have ever been strong enough to prod him into collecting abroad. Encouragement came from Hooker, however, prompting him to visit islands Darwin had seen six years earlier.

In March 1842 Hooker received word from the Navy that a vessel was going to map the Azores, and it could take along a naturalist. He wrote to Watson and asked if he were interested. Watson replied that he was, changed his mind the next day because of a family obligation, but two days later had resolved his impediment and accepted the offer.[13]

Captain Alexander Thomas Emeric Vidal came from a naval family. His father served as secretary to three successive admirals, and his two brothers were also naval officers. He entered the Navy in 1803, and became a captain in 1825. Most of

---

[12] While editing *The Phrenological Journal*, he only published two short botanical notes, both relating to the effects of the low temperature in January 1838 upon vegetation around his home; Watson 1836h, 1837h, 1838c and 1838d.

[13] Watson to Hooker (24 March 1842): 'valuable family property is held for the duration of 2 lives; one life, being my own, the other that of a sister, wife of Major Spence of the 31st, & unfortunately at present she is with the Regiment, which is serving in India', and to hazard both out of the country at the same time seemed irresponsible. Two days later he received word that Louisa Judith was on her way home, and would presumably arrive before the ship left for the Azores.

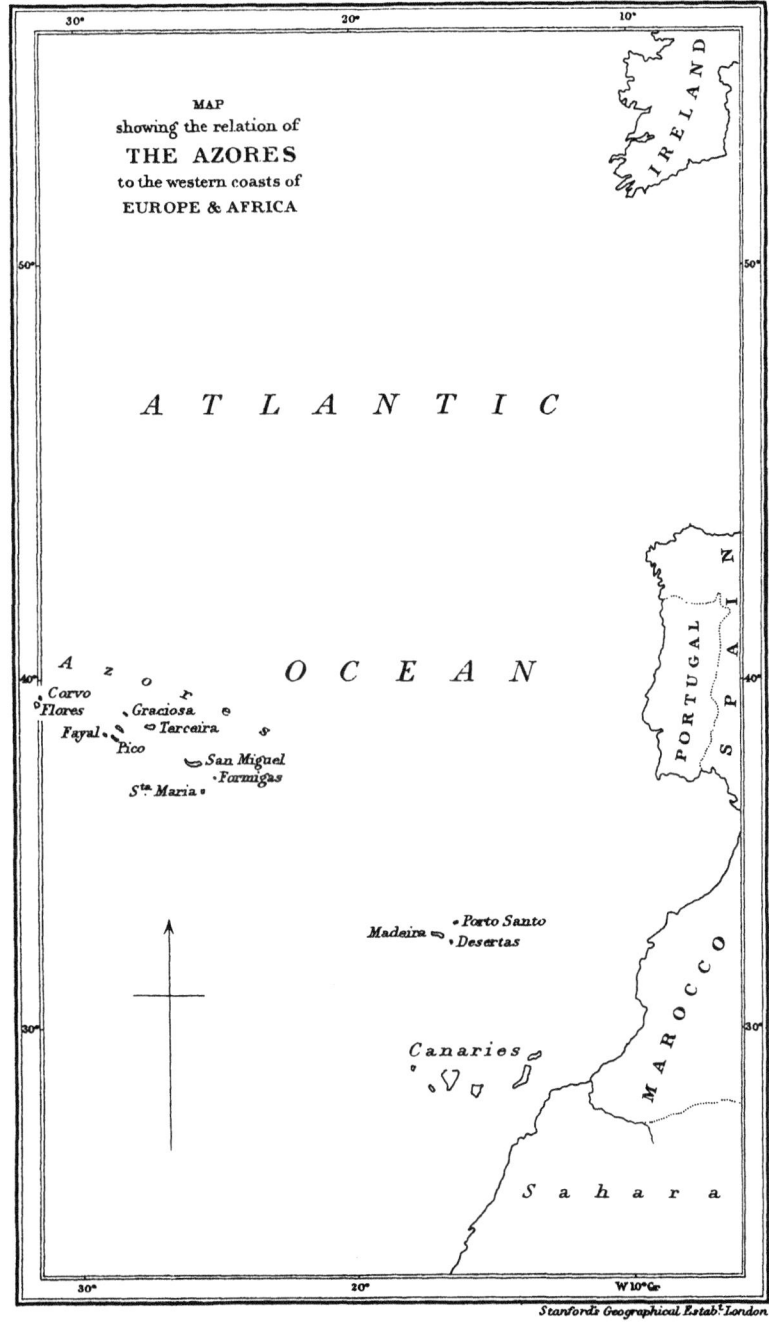

Figure 5 Locator map of the Azores Islands. Godman (1870).

his career was devoted to surveying the coasts of Africa and other regions. In December 1831, being between assignments and in Plymouth, he dined with a fellow surveying captain, Robert FitzRoy, and his naturalist companion, Charles Darwin, while they awaited a favorable wind to take their expedition to South America. Captain Vidal was an expert when he was assigned to map the Azores.[14]

Watson's preparations proceeded slowly; they probably included writing his will. He wanted to come up to Kew and borrow materials from Philip Barker Webb – specimens and/or volumes – on the Canary Islands which Hooker had offered to lend him.[15] However, it became difficult for him to find the time (24 April): 'each successive day I have been called either to Woolwich or London, or been detained by friends coming hither with the well-meant intention (but inconvenient under the circumstances) of spending a day with me before my departure'. Vidal's ship, the *Styx*, needed carpentry and painting and would not leave until May. Watson apparently went to Kew before leaving, because he saw Hooker's parcel of plants from the Azores which the Swiss botanist Guthnick had collected in 1838.[16]

On 3 May 1842, Watson went aboard, and some officers told him they would leave the next day. He found that inconvenient and complained to Hooker (4 May) that his cabin was not even ready to receive its furniture. However, they did not leave until 18 May. Like Darwin on the much longer voyage of the Beagle, Watson had to pay his own expenses, and he did not enjoy the voyage. He discovered that 'War Steamers' were not designed for passenger comfort.[17] Fortunately, they traveled at a brisk rate and arrived early on the 25th. Watson was thrilled by the sight of:

the lofty Peak of Pico, rising high and sharp into the deep blue sky, with a wreath of white clouds floating like a loose drapery around its dark sides, much below the summit.

Before 1 p.m. they dropped anchor at Horta, the principal town of the island of Fayal.

---

[14] O'Byrne 1849, 1229. Stone 1980, 140.

[15] A copy of Watson's undated will from about this time is now in the Archives, Royal Botanic Gardens, Kew. (The original is a mounting paper for one of his specimens of bottongrass.) See Webb & Berthelot 1835–50 and Desmond 1977, 648. Much later, Watson compared their monograph unfavorably to another: 'The *Manual Flora of Madeira* [1868], by Mr. [Richard T.] Lowe, is much the best authority for that island, so far as yet published, Ranunculaceae to Compositae in part. Webb and Berthelot's more ambitious work on the Canaries, although a valuable contribution to the literature of natural science, is perhaps not equally reliable in its botanical details.' Watson 1870, 123.

[16] Guthnick was a native of Berne, Switzerland, and became director of its botanic garden; Guppy 1917, 362, and Pritzel 1871–77, 132.

[17] Watson 1843–47, quotation, 3; paying his own way, 1.

Watson's account of his trip shows that he could write narration and description that bears comparison with Humboldt's and Darwin's travel books. But both of these explorers had described five-year expeditions in which the discoveries and events had compensated for the hardships. Watson was in the Azores for only three months, and one wonders when reading about it whether the inconveniences did not seem more important to him than the rewards. Yet, if one discounts much of his complaining as a result of his temperament, one sees that he did have rewarding moments which he described effectively. The extinct crater of Fayal was 'as peaceful and lovely a scene as I ever beheld', and it contained 'a natural botanic garden, where the true Flora of the Azores, above the cultivated region, reigns undisturbed by plough or spade'.[18] The people did not particularly interest him. He did not mention whether he had any difficulty communicating with them, but his phrenological evaluation, sent to Combe (21 December 1842), was unfavorable. Azoreans, in comparison to the English, were deficient in the organs of both Wit and Ideality, but:

Self Esteem is decidedly large, and perhaps [is evident] in the impulse to lower others by nicknames. Secretiveness is usually large also.

Although plant geography and species variability were uppermost in Watson's mind, he followed Humboldt's tradition of describing the geography of domestic as well as wild species. Sometimes these descriptions contain what we now call ecological aspects. For example: 'Strawberries do not succeed well, and the fruit which they do bear is with difficulty preserved from the innumerable blackbirds.'[19]

The ascent of Pico was a challenge which few European explorers could resist, and Watson's stiff right knee did not inhibit him. The other climbers were Captain Vidal, Lieutenant Cleaveland, and Assistant Surgeon Speer. The canopy of clouds at the base of the mountains was so dense on 1 July that Vidal doubted he could make theodolite readings and wanted to postpone the trip. Native porters convinced him that the clouds would dissipate, however, and they set out at about 5 a.m. They had the use of 'two beasts of great rarity in Pico, namely, a pony and an ass, which ... we bestrode in turn while ascending the lower part of the hill'.[20] A great challenge for Watson was observing the influence of elevation on the composition of vegetation. The previous summer he had conducted a similar study in Scotland's Grampian Mountains, where changes in elevation, climate and vegetation were less extreme. When they set out, he found 'a few plants of *Hyoscyamus Canariensis*, being the

---

[18] Watson 1843–47, 4 and 127.
[19] Watson 1843–47, 5.
[20] Watson 1843–47, II, 396.

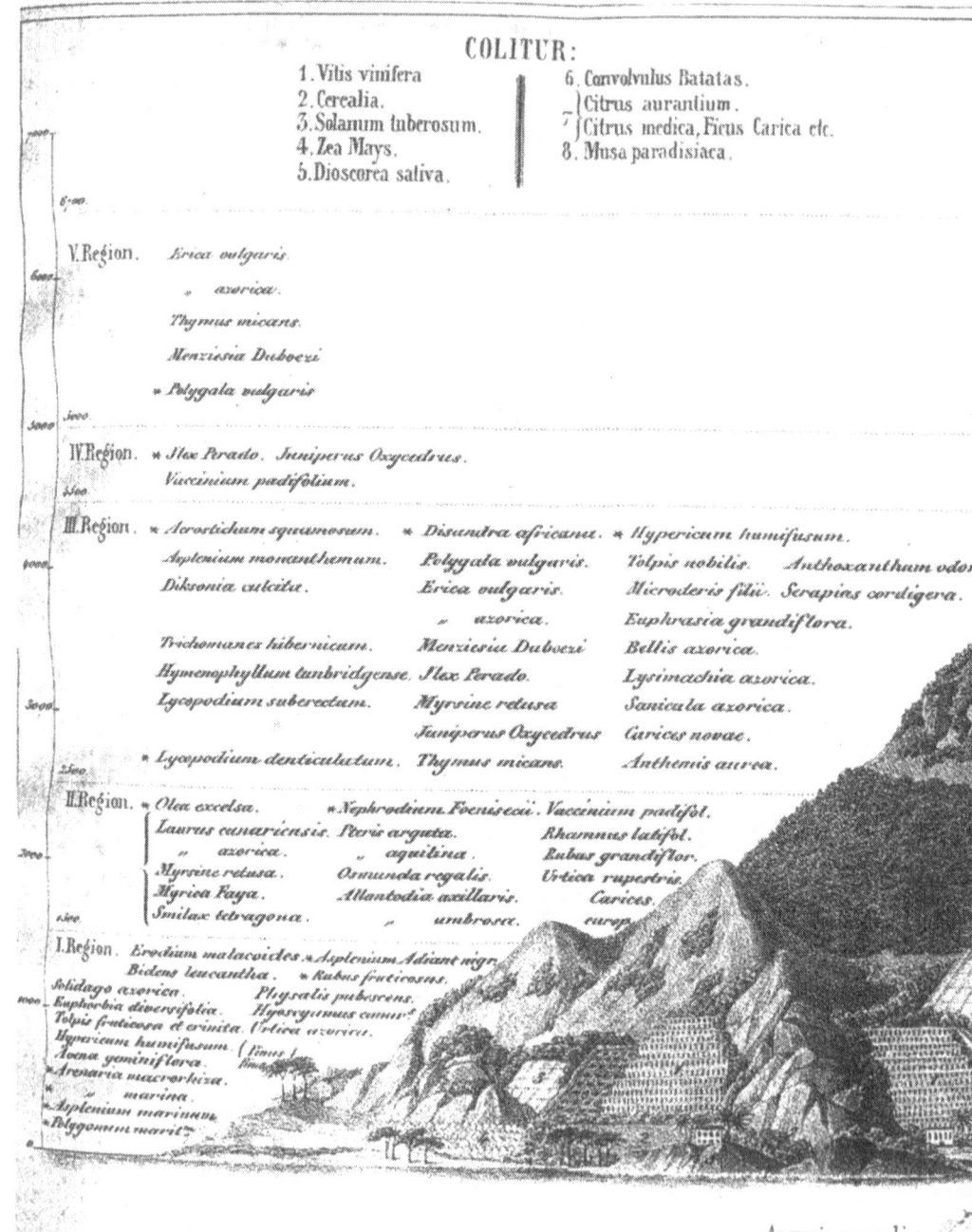

Figure 6 Diagram of Pico, by C.F. Hochstetter.

only spot in which it was found during my walks about the islands'.[21] This must have been close to shore, since the road went a short distance along shore before turning inland through 'an intermidable net-work of *vineyards*'. He noted down the kinds of weeds growing between cultivated plots and the road. He discovered, unexpectedly, that the fruits ripened here earlier than in Horta, and he decided it was because the town was built:

> on the south-east base of a range of fertile hills, and not ten miles distant. Probably the dark lava-rocks and walls of Pico, sparingly covered with vegetation, and thus often heated strongly by the rays of the sun, may be the chief cause of this peculiar result.

His observations on the distribution of domestic species were perceptive, though not extensive enough to distinguish (if that is possible) the cultural from environmental influences in determining distributions. At about 1000 feet elevation, 'the *orange* has disappeared; *fig* trees had become more numerous than below; and the *vines* were giving place to *apple* trees, of stunted size, and producing small fruit of little flavour, as I afterwards ascertained, for at this time the fruit was not full grown'. As they continued upward, yams (*Caladium*) and potatoes, 'which were scarcely seen lower down, indicated a transition from orchards to field crops. At first we saw occasional patches of these vegetables, intersperced with the fruit trees.' Still higher, 'indigenous shrubs took the place of planted fruit trees; single bushes or clumps of *Laurus* (*Canariensis* or *Barbasana*?), *Myrica Faya, Myrsine retusa, Erica scoparia* and *Juniperus* (*communis*?) being left to grow on stony or rocky spots that were unsuitable for the cultivation of the tuber-bearing vegetables'.[22] Cultivated plants disappeared as they continued climbing into a zone of shrubs, with other plants, whose relative frequencies he described, found in openings. Above the zone of mostly shrubs were clumps of shrubs interspersed within 'grassy swards' which contained:

> many small pools of stagnant water, which gave an abode to *Scirpus fluitans, Scripus Savii, Carex stellulata, Callitriche verna, Peplis Portula* and *Potamogeton natans*. Though very small and shallow, these pools are kept supplied with water by the mists and clouds from which this part of the mountain is seldom quite clear.

---

[21] Watson 1843–47, II, 396. For his catalog of Grampian plants in relation to altitude, see Watson 1842b.
[22] Watson 1843–47, II, 397–8.

He identified half a dozen species of grass, two of which were seldom seen below 1000 feet elevation.

As they emerged above *Erica scoparia*, Vidal thought they were above the limits of heath, but Watson noticed around them fronds of bracken fern (*Pteris aquilina*), and 'as that heath ascends in Scotland far above the *Pteris aquilina*, I read the appearances of the latter as a fair indication that we were still within the natural limits of heaths', and as they ascended to a less clouded atmosphere, between 4000 and 5000 feet, *Erica scoparia* reappeared. When Pico's summit came into view, they were again in a zone of evergreen shrubs, but tiny ones compared to the same species at lower elevations. The uppermost vegetation consisted of ling (*Calluna vulgaris*), a thyme (*Thymus caespititius*), a few mosses, and lichens interspersed among boulders, but the peak itself was practically bare.

Pico is an inactive volcano, with a large crater enclosing a smaller one. The explorers entered both, the ground within the smaller one being hot and steaming. Temperature at the peak was 53°F – 22° cooler than at sea level. Wind chill made the peak seem even colder. They had brought along too little cider and whisky for the six of them, and Watson's lips and tongue became parched as they descended to their baggage lower on the mountain side. Later, Vidal determined Pico's elevation barometrically as 7616 feet. Watson then estimated that *Calluna* and *Thymus* grew at 7000 feet, *Erica* reached 6000 feet, and the upper zone of other evergreen shrubs reached 5000–5500 feet. They spent that night within these latter shrubs. Watson doubted that the highest cultivated plants, cocoa and potato, grew above 2000 feet. He regretted not being able to determine exact elevations at which various species grew.[23]

Besides Fayal and Pico in Central Azores, the *Styx* visited Corvo and Flores in Western Azores, and Watson collected on all four islands. Although Vidal and his crew were somewhat helpful, Watson did not accomplish all he wanted. In 1843 he only explained his difficulties in drying plants – an uncharacteristic restraint – but in 1870 he explained more fully the problems he had encountered. The *Styx* had visited only four of nine islands, not including the larger ones of San Miguel and Terceira, and Watson's transportation for research was poorly accommodated: 'And when I took advantage of the boats of the island fishermen for this purpose, I lost much of the day through delays in the morning and through feeling compelled to be back on board the English vessel at night, lest she should have gone elsewhere, leaving me behind.' He acknowledged that Vidal was 'an able surveyor, an intelligent and agreeable man as an acquaintance on shore', but aboard ship he had a strange idea of discipline: he kept his officers and crew on their toes 'by keeping everybody in ignorance of his intentions, or the intended positions and movements of his ship from day to day'. However useful that may have been for discipline, for

---

[23] Watson 1843–47, 404–6. Vidal's map of Pico is reprinted in Guppy 1917, facing 359.

Watson's purposes it was very inconvenient and expensive, since he frequently had to hire boats to carry him from shore to ship, which was sometimes miles away. After returning, he had to dry his plants in his cabin, which was only about six feet by five feet.[24] Given Watson's tendency to complain, it is helpful that an independent source attested to his difficulties. H. David Graham Miller, MD, was surgeon on the *Styx* in 1842, and in 1846 he wrote a recommendation for Watson, stating:

> I had the amplest opportunities, from almost daily intercourse with him, of ascertaining the success of his labours; and I can truly affirm, that his industry was indefatigable, and his personal trouble very great, in investigating the botany of the Islands: – nothing short of such perseverance could have enabled him to accomplish so much as he did; the Islands, generally, being very rugged and mountainous, and many of the localities which he visited, of very difficult access.[25]

This frustrating trip came to a frustrating end. Characteristically, Watson's published account of the expedition begins with a description of his troubles getting home. He decided to return on a mail steamer:

> This resolution I was unfortunate enough to carry into effect, and by so doing, was subjected to the tediousness of a rough passage, protracted to twice its usual length, through sheer mismanagement in taking on board the mail-packet barely sufficient coals to carry us to Falmouth with a fair wind. The wind proved adverse the whole way, and for a few days blew a hard gale, so that our stock of coal was exhausted before we could make the English Channel; and there was no resource left but that of turning back and running before the wind, under such small sails as could be raised in the steamer, across the Bay of Biscay to Corunna, for a fresh supply of coal. In this dilemma, it was some consolation to anticipate a botanical day or two on Spanish ground; but scarcely was our anchor down before he had notice from the Spanish authorities that none of us could leave the ship, which must be put under quarantine, in consequence of having come from the West Indies ... By coming home in the mail-packet Dee, I thus lost the opportunity of autumnal botanizing in Fayal, and merely wasted the time in playing at 'pitch and toss' in the Bay of Biscay.

Nor was that all. Shortly after reaching home, he complained to William Hooker (12 October 1842) that his herbarium sheets of specimens had been left aboard the *Styx*,

---

[24] Watson 1870, 115.
[25] Quoted in Watson 1846a, 32.

and that a case of empty paper had been sent on the Dee instead. He feared that the specimens would only arrive in November or December, in poor condition, 'since many of them were by no means thoroughly dry'.

Watson may have retrieved his specimens before December, for he sent off the first part of 'Notes of a Botanical Tour in the Western Azores' to Hooker in November, and it became the lead article in the 1843 *London Journal of Botany*. Coincidentally, the lead article in the 1843 *Archiv für Naturgeschichte* (Berlin) was 'Übersicht der Flora der azorischen Inseln' by Moritz Seubert and Christian Friedrich Hochstetter. Seubert published a longer *Flora Azorica* the next year, which Watson obtained after publishing the first three installments of his 'Botanical Tour'. He probably assumed that the book included everything from the article, but Hochstetter's marvelous sketch of the vegetation on Pico was not reprinted. Not only would Watson have found it very interesting, but its indications of elevations at which the different species grow might have caused him to take more seriously than he did Seubert's data on altitude distributions. Henry B. Guppy checked Hochstetter's data about seventy years later, and found them satisfactory.[26]

Although Watson commented to Hooker (12 October 1842 and undated, no. 263) that the Azores have fewer species than he had expected ('scarcely 300'), given its wide range of elevations and climates, there were nevertheless species from about a dozen genera which appeared to be unknown elsewhere.[27] The closing arguments for his third installment were intentionally provocative, if also pragmatic:

> The shrub which I have called *Vaccinium Maderense* is certainly the *V. cylindraceum* of Smith; but I cannot regard it as being specifically distinct from *V. Maderense*, of which, however, it is a very handsome variety, with flowers more numerous, and often twice the size of those in the Madeira specimens. Those botanists who delight in multiplying species on paper, by describing extreme forms, in disregard of intermediate and connecting links, will doubtless keep *V. Maderense* and *V. cylindraceum* distinct.

---

[26] Guppy 1917, 362–3. Seubert's book was based mainly on collections made by C.F. Hochstetter and his son, Charles, but he also had specimens collected by Guthnick. On C.F. Hochstetter (1787–1860), see Stafleu & Cowan 1976–92, II, 220–2.

[27] 'The genera of Fayal plants, which yield species that I have not yet been able to refer to described species are *Convolvulus, Carex, Euphorbia, Luzula, Veronica,* and *Rubus*. There are also species of *Carex, Cardamine, Bellis, Festuca, Sanicula* and *Lysimachia*, which have been named, if not described, by Lowe, Guthnic[k], or other botanists.' Watson 1843–47, 131. This statement is at the end of his second installment, published 1 March 1843.

Figure 7 *Campanula vidalii* H.C. Watson. W. Hooker (1844).

Much more to the same effect concerns questionable pairs of species or varieties within five other genera.[28]

Guthnick had visited only San Miguel, Terceira and Fayal, but after his departure the Hochstetters continued on to Pico, Flores and Corvo. Thus, they visited the four islands where Watson collected, and also Terceira and San Miguel, where he did not. Nevertheless, Watson had found species unlisted in Seubert's *Flora*, and therefore he could produce a more comprehensive list. Before doing so, he sought the expert opinions of Philip Barker Webb and Charles M. Lemann about the extent of similarity between various plants of the Azores and those which they had collected respectively in the Canaries and Madeira. Soon after Watson's return, Webb had proposed that they swap specimens, and Watson agreed. Unlike Lemann, however, Webb failed to provide the evaluations Watson sought. When publishing his list of species in 1844, Watson thanked Lemann, and lamented Webb's lack of assistance.[29]

Watson's list more or less superseded Seubert's, yet he remained dissatisfied with its scope. T. Carew Hunt, British Consul for the Azores from 1839 to 1848, lived on San Miguel; he volunteered to send plants from that island to Watson. They apparently never met, and perhaps Hunt wrote after seeing Watson's articles on plants of the Azores. Although Watson referred to him in 1870 as a botanist, Hunt is only known to have published geographical and geological descriptions of San Miguel and Santa Maria.[30] Watson happily acknowledged that 'The zealous and intelligent exertions of Mr. Hunt in the years 1844–48 conduced much to the increase and to the diffusion of knowledge about the botany of the Isles', and his collections provided the basis for Watson's 'Supplementary Notes on the Botany of the Azores'.[31] In exchange, Watson probably sent Hunt books and other information (probably including Watson's own copy of the *Vestiges of the Natural History of Creation*, mentioned to Combe, 14 May 1847). Yet when Hunt left the Azores in 1848, Watson never heard from him again.

Watson was not entirely dependent upon Hunt for his further researches on Azores plants. He brought back with him in 1842 seeds of some species, some of which he planted in his garden and others he sent to Hooker (4 and 25 January 1843)

---

[28] Watson 1843–47, 407–8 (published 1 August 1843). The other five discussed were *Juniperus communis, J. nana, Bellis Azorica, B. perennis, Myosotis maritima, M. arvensis, Luzula maxima, L. elegans, L. pilosa, L. Azorica*, and *Erythraea diffusa* varieties.

[29] Watson 1843–47, 584–5. On Webb, see above, n. 3. Charles Morgan Lemann (1806–52) collected in Madeira in 1837–38; Desmond 1977, 382.

[30] Hunt 1845a and 1845b. He cited Watson 1843–47 on 281 and 296. Watson mentioned Hunt's articles but never saw them. On T.C. Hunt (d. 1886) see Desmond 1977, 330, and Watson 1843–47, 116–17.

[31] Quotation from Watson 1870, 117. 'Supplementary Notes' is in Watson 1843–47, 380–97.

to plant at Kew. He also borrowed herbarium specimens from Hooker, Babington and others for comparison with Azorean specimens. In return, he provided sets of dried Azorean plants for the Royal Botanic Garden, Kew, for Babington (15 and 18 March 1843), and on loan to others (for example, unaddressed letter, 13 April 1846, in B.M.).

Hooker for many years published installments of *Icones Plantarum; or Figures, with Brief Descriptive Characters and Remarks, of New or Rare Plants, Selected from the Author's Herbarium*, and he invited Watson to select one of his new Azorean species for inclusion. Watson suggested (26 April 1844) that his Campanula would be a good choice because of 'its peculiar habit; & being a Flores plant, it is not likely to be among those in the hands of Hochstetter'. However, he thought that a good drawing would necessitate 'combining the peculiarities of the few specimens, rather than by giving a portrait of one only'. Although he lacked a draftsman's skill, the rough sketches, detailed explanations and herbarium specimens which he sent provided the basis for the illustration Hooker published of *Campanula Vidalii*, discovered by Captain Vidal 'on an insulated rock off the east coast of Flores, between Santa Cruz and Ponta Delgada'.[32]

For a while Watson thought of publishing a small book on the botany of the Azores (to Hooker, 4 January 1843), but gradually became engrossed in studies on the varieties and distributions of British species. Since he abandoned the idea of an Azores book, in the first and only installment of the very ambitious third edition of his *Geographical Distribution of British Plants* (1843, iv + 259 pp.) he indicated which British species were also found in the Azores. However, he dropped the discussions of extra-British ranges of species in *Cybele Britannica* (1847–59) to better insure that he would finish the work. Another reason for not writing a book on Azorean plants was the adequacy of Seubert's *Flora Azorica*, even if Watson did not accept Seubert's statistics on the number of species unique to these islands versus the numbers Azores shared with Madeira, the Canaries, Africa, western Europe, and the British Isles. Watson suggested that Seubert's book might provide a basis for testing the validity of Forbes's hypothesis about the migration of floras across land bridges.[33]

---

[32] Hooker 1844, Tab. 684.
[33] [Watson] 1846h.

Chapter 7

Relationship with William Hooker, 1834–50

> You take with sufficient good temper the circumstance of a botanist siding with the Enemy.
> Watson to W. Hooker, 25 January 1849

William Jackson Hooker (1785–1865) was a leader of British botany for five decades, and for about half that time he and Watson were regular correspondents. Hooker deserves much credit for encouraging Watson's studies, and Watson responded initially with much gratitude and little criticism.

Hooker was a son of a merchant's clerk, and his formal education prepared him for a business career. His interest in botany led to his being introduced in 1804 to banker and amateur cryptogamist Dawson Turner, who brought him into his family's brewery and in 1815 gave him a daughter in marriage. In 1809 Turner introduced him to Sir Joseph Banks, who arranged for him to botanize in Iceland and later attempted unsuccessfully to arrange for him to do so in Ceylon. In 1820 Banks assisted him to obtain the Regius Professorship of Botany at the University of Glasgow, where his popularity and industry were eventually witnessed by Watson. Hooker published extensively throughout his career. He left the University in 1841 to become the first Director of the Royal Botanic Gardens at Kew. There he remained, steadily expanding and improving its gardens, herbarium and library. His son Joseph became his assistant in 1855 and later his successor. Both father and son were knighted for their contributions to botany and empire.[1]

Watson's travels and collecting in England, Scotland, and Wales would never provide enough specimens to meet his research needs, and he decided to sacrifice an earlier hobby for a present need. As late as 1830 he had mentioned to Hooker that he bought insects as well as plants, but on 10 July 1834 he offered to swap his insect collection with Hooker's son, Joseph, in exchange for Scottish plants. William Hooker feared that Watson was cheating himself, but Watson disagreed and explained further his wishes (13 August 1834):

---

[1] Allan 1967, chaps 1–14. Desmond 1977, 319 on W.J. Hooker, and 622 on Turner; Allan 1995, chaps 10–16. Hooker 1903. Stafleu & Cowan 1976–92, II, 283–301.

> My object is to collect together specimens from as many counties as possible, of all save the commonest plants; these latter I should like to have from several remote stations, (east, west, north, south, low & high ground) but do not care to have them literally from every county. I have very few common plants from the north or west of Scotland.

Joseph fulfilled his part of the bargain, yet Watson's suspicious nature eventually led him to devalue Joseph's specimens. He confided to Scottish botanist John H. Balfour (30 December 1844):

> I possess scarce any specimens, from what I call the 'West Highlands', excepting some Dumbarton (chiefly) & Argyle specimens collected by Dr. Joseph Hooker, while yet a boy, & I fear that he then mingled his specimens from different counties. I would much prefer to have your name to cite as the authority for the Western localities of specimens.

It is unclear that his doubts about Hooker's collection were based upon anything more substantial than the fact that he had been young when he made it, and by 1844 Watson might even have forgotten that Joseph had been about 17 years old.

The affable William Hooker expressed a hope of seeing Watson in the autumn – perhaps at the meeting of the Naturalists' Association – but the latter declined (27 August 1834): 'The love of solitude makes me shun such scenes.' This response was not an excuse to avoid Hooker personally, since in the following year Watson dedicated to him *The New Botanist's Guide to the Localities of the Rarer Plants of Britain*. Watson also continued to purchase from Hooker Thomas Drummond's American plant specimens, but on 8 February 1836 he sent a final payment and expressed concern about supporting the widow of Drummond, who had died in Havana.[2]

As Watson prepared to give up editing *The Phrenological Journal*, he may have coveted the impending vacancy at the Royal Botanic Gardens. According to one gossip, Watson had had his eye on the directorship when he settled at Thames Ditton in 1834,[3] which was also the year in which Hooker began seeking support for himself as a candidate. However, there is no evidence that Watson ever wrote to Hooker asking for support in his application for the position, as John Lindley did in

---

[2] On Thomas Drummond (c.1790–1835) and his collections, see Coats 1969, 314–6, 325–7, Desmond 1977, 196–7, and Stafleu & Cowan 1976–92, I, 685.

[3] Thomas Bruges Flower (1817–99), as related to William Bowles Barrett (1833–1915) in 1882. MS notebook at Weymouth Public Library, p. 20. I am indebted to David E. Allen for this information. On Flower and Barrett, see Desmond 1977, 227 and 41.

1838.[4] Perhaps Watson learned that Hooker was the strongest contender. Even though Watson seems to have realized he had no chance of obtaining the post, frustration at not obtaining the directorship may be evident in his occasional complaints to Hooker or to others about how Hooker ran the Gardens. On 18 November 1841 Watson accepted Hooker's invitation to visit Kew, but declined to have dinner with the Hookers: 'I will beg to limit my visit to the morning; and in so doing, shall feel that I am trespassing quite enough on your time.' He probably gave Hooker little time to show off his new domains, because he took the opportunity to show him some of his own herbarium specimens which Watson felt provided evidence for revisions in the pending fifth edition of Hooker's *British Flora* (published 1842).

Although botany was not a suspect science like phrenology, Watson nevertheless considered the best means for encouraging its progress. In 1836 his concern had been with raising standards (6 January):

... your introduction of 'Observations on British Plants' into the Comp[anion] to Bot[anical] Mag[azine] ... will incite observers to make exact observations; and it will also, if continued, supply the place of a repertory for small points not singly of sufficient importance to swell into separate papers. In fact, we want a periodical for such matters, Loudon's Magazine of Natural History being framed too much on the popular science plan to admit of this properly.

Five years later, Watson advocated to Hooker a democratic rather than an élitist model for the spread of botany, as he had somewhat earlier advocated for phrenology with George Combe. The cases were similar in a crucial respect: Hooker's *Journal of Botany* had a small circulation, as had *The Phrenological Journal*. Watson advised (16 November 1841) that the *Journal of Botany* needed more variety in order to expand its circulation. Illustrations and descriptions might be indispensable to systematic botanists,' but only a small number of botanists are sufficiently advanced in their science to take much interest in descriptions of plants which they are not likely to see. – That small number alone cannot adequately support a monthly Journal.' Secondly, Hooker's Journal should not be '*too far in advance of readers*. In botany, as in everything else, we find a greater demand for the commonplace than for the excellent.' Perhaps Hooker had suggested that Watson contribute an article, for Watson agreed to send one on the range of species in the Scottish Highlands, 'in regard to height, somewhat after the plan of [Augustin] De Candolle, in his paper in *Mémoirs de Physique & de Chemie de la Société*

---

[4] Allan 1967, 96. On W. Hooker's tenure at Kew see also Blunt 1978, chaps 11–20; Hepper 1982, 16 *et passim*; Turrill 1959, 26–9.

Figure 8 William Jackson Hooker (1841), by David Macnee.

d'Arcueil.[5] Watson's 'Catalogue of Plants of the Grampians, Viewed in Their Relations to Altitude' was a novelty in a British journal when published in Hooker's *London Journal of Botany* in 1842.[6]

Watson also urged Hooker to establish a garden of British plants at Kew (3 March 1842) since 'the seeds, at least, of many [species] might be readily obtained through the Botanical Societies, or individual botanists. The post now affords great facility for sending roots; enclosed in tin boxes.' Watson had begun such a garden for himself but could not afford to complete it. This advice was apparently unsolicited and was not implemented, but that lack of action did not reflect a lack of respect for

---

[5] Candolle 1817.

[6] Confusingly, Hooker's *Journal of Botany*, about which their discussion began, was terminated in 1842 (after vol. 4) and Hooker then started his *London Journal of Botany* (vol. 1, 1842) in which Watson published two articles. To unravel such journal complexities, see Lawrence et al. 1968.

Watson's judgment. Twice Watson attempted to obtain university appointments as a botanist, and twice Hooker supported him – perhaps his most important support.[7]

It was after Watson failed to obtain a professorship at King's College, London in 1842 that Hooker provided an opportunity for him to divert his mind from it with a botanical trip to the Azores. Since that trip was successful, in early November Hooker asked if he would also like to go and study the plants of the Canary Islands. Watson replied (5 November 1842) that he would, but not next year; he never went. Hooker mentioned in April 1846 that there was an opportunity for a botanist to establish a botanical garden at Cape Town, South Africa. Hooker was glad to support Watson's interest in the position (Watson's reply, 26 April 1846), but then complications led him to abandon the idea. Three months later he decided to apply for a chair of botany in Ireland, and Hooker not only wrote another letter for him, but also offered to speak on his behalf to the Archbishop of Dublin.

Joseph Dalton Hooker, discussed in passing until now, warrants more detailed attention here. He was born in 1817, and developed an affectionate relationship with his father. In 1832, William Hooker wrote to Joseph's grandfather, Dawson Turner, that Joseph, then about 15 years old:

> is already a fair British botanist and has a tolerable herbarium, very much of his own collecting. But the orchidae are his great favourites, and he has an eye for them, and a memory too for their names, which often surprises me. Had he time for it he would already be more useful to me than Mr. Klotzsch [his assistant].[8]

Four years later, Joseph's mother wrote (7 May 1836) to her father, the same Dawson Turner, that:

> A fortnight ago, Joseph walked 24 miles – from Helensburgh to Glasgow – rather than wait for the steamer next morning, by which delay he would have missed a lecture ... Joseph has been equally earnestly employed in turning over stones and hunting in the rejectament of the sea for beetles. His collection of insects is becoming considerable, he devotes every spare minute to it, and has opened a correspondence with several entomologists, both British and foreign.

She proudly announced that, although the youngest student in his class at the University of Glasgow, he had won prizes in his courses on natural philosophy and anatomy. Joseph's parents obviously believed that he could collect plants for Watson without mixing up their counties.

---

[7] W. Hooker, 17 January 1842, quoted in Watson 1842a, 7.
[8] Quoted in Huxley 1918, I, 25, with information also on S.J. Klotzsch.

In autumn 1838, some eight months before receiving his MD degree, Joseph was asked to be the assistant surgeon and botanist on Captain James Clark Ross's Arctic expedition, which was expected to leave about the time Joseph received his degree (though its actual departure was not until 30 September 1839). The expedition lasted four years, and Hooker proved to be as diligent a field botanist as he had been a student. His collections and observations provided the basis for his important contribution to phytogeography, *The Botany of the Antarctic Voyage of H.M. Discovery Ships Erebus and Terror in the Years 1839–43*.[9]

Upon returning, Joseph wanted a permanent position. Robert Graham was ill in late 1844, and on Robert Brown's recommendation he invited Joseph to lecture for him in the spring term of 1845. Graham died the following August, and Joseph became a candidate for the chair. By the time that Joseph asked him for a recommendation, Watson had already written one for George Dickie, but Watson assured Joseph (26 August [1845]) that he was better qualified, and he wrote a strong recommendation that praised *The Botany of the Antarctic Voyage,* and then said:

Undoubtedly the name of Dr. J.D. Hooker will descend to future ages, with the honour which is awarded to those who carry forward scientific knowledge beyond the state in which they find it, – the surest test of personal preeminence.

Joseph also had the backing (among others) of the Crown. However, J.H. Balfour, who had succeeded William Hooker as Professor of Botany at Glasgow, and who was both a native of Edinburgh and a graduate of its medical school, had stronger backing in Edinburgh, and the chair went to him. Joseph thought that he was better qualified than Balfour and Watson agreed (to Joseph 12 December [1845]).

A short time later, Henry de la Beche, Director of the Geological Survey of Great Britain, wanted to hire a botanist, and consulted William Hooker.[10] Because Watson was interested in the situation – primarily, he claimed, as a taxpayer who opposed the misallocation of public funds – and brought it up on several occasions (while having little information), it is worth quoting from the letter which Joseph Hooker wrote to Sir James Ross when accepting the post:

The duties will leave me more than enough of time to carry on my [Antarctic] Flora as fast as the plates can possibly appear, but I do not know what the Admiralty will say to my taking the duty. My work has in many ways cost me

---

[9] The six volumes appeared in three parts of two volumes each: *Flora Antarctica* (1844–47), *Flora Novae Zelandiae* (1853–55), and *Flora Tasmaniae* (1860). For modern extracts, see Hooker 1953, chap. 8.

[10] Huxley 1918, I, 207.

already nearly £100, and I believe I have never made 6d. by it and never shall. If the new duty were to interfere with my Flora, or were my salary so good as to make me independent of the Admiralty, I should not think about drawing any further Admiralty pay, but as that is not the case and as I have never made a farthing by my Botany work, I think of making a push for the continuance of my pay when I enter upon my new duties. I should feel very obliged for your opinion of how their Lordships are likely to regard my views.[11]

None of the parties to this understanding had any reason to inform Watson of its terms. Yet he resented Joseph getting the position and expressed his feelings to Charles Babington, a botanist with whom he already had a strained relationship:

What say you to Dr. J.D. H[ooker]'s appointment as 'botanist to the Geological Survey'? I do think this *rather* worse judgment than preferring Balfour to Dr. H. in Edinburgh. *He is a botanist*; but his acquaintance with British plants must be extremely small; & he is paid (by a sinecure navy appointment, at 1000£) to devote his time to the Antarctic botany. If 150£ a year was to be given to British botany, it should have been given to somebody who knows British plants better, & would devote time to *British* botany. Were I not a personal friend, or something like it, I would get attention called to the appointment, by an M.P. in the House of Commons. I suppose Mr Forbes managed the matter, wanting somebody to help him 'to pursue the subject' of five British floras, 'in connexion with the Geological Survey'.[12]

Although Edward Forbes may have participated in the Geological Survey's discussion to hire a botanist, it was the Director who had gone to William Hooker for advice. Watson's reference to Forbes's 'five British floras' was a sarcastic allusion to the phytogeographical subdivisions which Forbes had proposed for Britain at the previous year's British Association meeting.

When Joseph heard of Watson's grumblings, he wrote to him, and Watson responded (7 April 1846):

you cannot suppose me to doubt your individual qualifications; but I looked upon you as already engaged (purchased at public expense) to devote your time & attention to Antarctic botany; while there were many others who are competent to look after British botany ...

---

[11] Not dated, but probably written in February 1846. Quoted in Huxley 1918, I, 208–9.
[12] This letter is only dated '1846', but since Watson asked Babington about the recent meeting of the British Association for the Advancement of Science, held at the end of August 1845, he probably wrote it in early 1846.

Figure 9 Joseph Dalton Hooker (1855), by George Richmond.

Joseph Hooker worked for the Geological Survey from February 1846 until October 1847, and made creditable contributions to paleobotany.[13]

Watson could not afford to feel hostile toward either Hooker for long, because in July 1846 he learned that the government would establish two new colleges in Ireland, with chairs of botany. He requested from father and son letters in support of his application for one of them.[14] They complied, then five months elapsed without further correspondence between Watson and William Hooker. Perhaps fearing that his complaints about Joseph had been relayed from Babington to the Hookers, Watson wrote to William on 2 March 1847 to ask if he had inadvertently caused offense. William assured him that he had not, and their correspondence about the identity of plant specimens resumed.

Another issue Watson raised which probably annoyed the Hookers was Joseph's plan for a botanical expedition to India in 1847. Watson wrote to each of them expressing the opinion that it was wrong to go on another expedition without first completing the publications and other tasks specified in his paid contract from the Admiralty.[15] At this, the Hookers may have wished Watson would mind his own business, but they could not say so, because a government contract was every citizen's business. Watson explained to Joseph that if he died on the new expedition, his unpublished findings from the previous one would be lost to science. After Joseph had long departed to India, Watson wrote to William in a more conciliatory tone (18 December 1849): 'his time is quite as importantly devoted to science, tho' in a different field'. When Watson read in the newspapers about Joseph's detention in Sikkim, he commented to William: 'Doubtless he would soon be released from durance; though it would seem that Dr Campbell had behaved improperly & injudiciously'.[16] Historians have not supported Watson's suspicion, and William doubtless depended upon others rather than Watson for consolation. About two-and-a-half months later William dampened Watson's complaints about Joseph by explaining the contents of his contract with the Admiralty. Consequently, Watson retreated:

From your letter of today, I guess that there is an erroneous belief current among botanists, touching the arrangement or contract between Dr. Hooker & the Admiralty ... Facts of course destroy errors in time; though silence leaves them

---

[13] Turrill 1963, chap. 3.
[14] He wrote to W. Hooker on 28 July and to J. Hooker on 5 August.
[15] Watson to J. Hooker, 9 October 1847; to W. Hooker, 18 December 1849.
[16] Watson to Hooker, 26 January 1850. Joseph and Archibald Campbell had already been released by that date. Allan 1967, 176. Huxley 1918, chap. 15. Turrill 1963, 70.

Figure 10 Drawing by Ella Taylor: Princess Mary giving instructions to Craig, a gardener, Royal Botanic Gardens, Kew, Oct. 29, 1858, with Sir William Hooker and Ella Taylor. This is the image of Hooker that Watson held in Hooker's later years.

to spread. I begin to suspect that I am myself under mistake, equally with various others.[17]

Hooker may have accepted this quasi-apology, but whenever one issue was resolved, another was likely to appear.

Watson informed W. Hooker (15 January 1849) that he had taken a political stand against unnecessary government expenditures, though his strictures were not aimed at Kew. Hooker seems to have assured him that he was not offended by either his

---

[17] Watson to Hooker, April 1850. This letter is numbered 360 in vol. 28 of Hooker's correspondence. Assuming they are numbered as received, it was written before 13 April, which is the date of Watson's letter, no. 361.

politics or the policy derived from it. Watson explained his position further on 25 January 1849, but in doing so, he undermined his previous claim of not targeting extravagance at Kew:

You take with sufficient good temper the circumstance of a botanist siding with the Enemy. I look at government expenditure through the glass of political economy, & cannot justify outlay for ornament or unproductive science while exorbitant taxation is hurrying the nation to insolvency or revolution, or both together.

The Palm House, completed in 1848, had taken three-and-a-half years to build and was the largest and most expensive greenhouse in the world.[18] It was the central attraction at the Royal Gardens and a monument to Hooker's vision and leadership. Oddly, despite his practical perspective, Watson seemed uninterested in the Museum of Economic Botany which Hooker had opened at Kew in 1847. Hooker was an early example in Britain of a civil servant who was a science administrator – something Watson observed but little appreciated. He was more concerned about the waste of tax funds than he was about government support for science. A modern study on the importance of the Royal Botanic Gardens at Kew for British colonial agriculture, shows the British taxpayers received an excellent return on their investments in these Gardens.[19] Although that outcome might not have been easily discerned in 1849, Watson was not giving his friend, Sir William, the benefit of any doubts.

Nevertheless, Hooker requested that Watson review Jules Thurmann's *Essai de phytostatistique* (1849) for the London Journal of Botany. Watson had already received a copy of the book from its author and agreed to do so, using the occasion to berate his peers – including William Hooker:

English botanists can repeat, or copy, or imitate very successfully, but they do not often originate new views and improved modes; they do not discover and invent in botany. They can detect and observe, can depict and describe, can catalogue and group individual facts and objects; but they do not connect their data, thus acquired, by the theoretic relations of cause and effect, nor do they often generalize their details, so far as to elevate these into the category of scientific principles.

With certain reservations, the essay of M. Thurmann may be cited as an example of the opposite kind.[20]

---

[18] Allan 1967, 147–51. Blunt 1978, chap. 13. Desmond 1995, 160–6. Hepper 1982, 4, 17, 22, 60–3.

[19] Brockway 1979.

[20] Watson 1850a, 187–8. Watson sent his review to Hooker on 2 May. Watson also published his translation of Thurmann's summary of his *Essai*; see Thurmann 1850.

A decade earlier Watson had berated phrenologists in a similar tone, yet had done very little to set an example for them. He did, however, set an example for the botanists.

Later, Watson wrote (6 October 1850) expressing hope that W. Hooker had recovered from an illness and thanking him for a copy of the sixth edition of *The British Flora*. It was co-authored for the first time with Balfour's successor in the chair at Glasgow, George Arnott Walker Arnott (1799–1868), who was also Hooker's long-time friend and collaborator.[21] The compliment which Watson offered – 'Dr Arnott has evidently bestowed considerable care & attention in preparing this edition' – was accompanied by a complaint:

> Likely enough, the long note on page 570 is from your pen. – If so, allow me to say that Mr Babington took the name of *[Lastrea] foenisecii* from me, & passed it off as his own first application: it was a theft in science, for which he got reproved in *Phytologist* ii, p. 877.[22]

Watson's complaint must not have impressed Hooker and Arnott, because the seventh edition also credited *foenisecii* to Babington.[23]

---

[21] Desmond 1977, 19. Stafleu & Cowan 1976–92, I, 67–9.

[22] Watson to Hooker, 6 October 1850. The note in Hooker & Arnott 1850, 570, is for the account of *Aspidium spinulosum* Willd. and states, in part: 'On the *quaestio vexata* of the differences, specific or otherwise, between *A. spinulosum* and what are here considered varieties, we have little to add, although we have not failed to reconsider the subject fully ... One state of the plant, however, we are here desirous to notice, from the great discussion it has occasioned in some of the periodical journals, namely, *Aspidium dilatatum*, var. *recurvum*, of Bree in *Loud. Mag. of Nat. Hist.*, vol. 4, 163. cum ic.: *Lastrea recurva* of Mr. Newman in *British Ferns*, 1844, 226. We find no specific character in the latter work; but this deficiency is compensated by Mr. Babington, who (*Man. of Brit. Botany*, ed. 2, 411), under the name *Lastrea Foenisecii*, thus distinguishes it [Latin quotation].'

[23] Hooker & Arnott 1855, 586. Watson's later comments on the sixth edition are quoted on pages 132–3.

Chapter 8

Seeking Employment, 1842–48

... you are eminently qualified to fulfill the important duties of the Chair in question.
William J. Hooker to Watson, 17 January 1842

When the regular lecturer withdrew in the summer of 1837, Watson taught a ten-week course on botany to medical students in Liverpool.[1] That might have been a first step toward a permanent teaching position if he had not been leaving botany for phrenology. After switching to phrenology, he would not have contemplated seeking a teaching position, because British universities never hired professors of phrenology. In 1841 Watson resumed his correspondence with botanists by congratulating J. Hutton Balfour on becoming successor to William Hooker as Professor of Botany at Glasgow (11 April).

Watson visited Hooker at Kew in November, and they undoubtedly discussed Watson's trip of the previous summer to study the range of species in the Grampian Mountains of Scotland. Watson soon sent him a paper on that subject for publication in *The London Journal of Botany*, and also detailed suggestions for a new edition of Hooker's *British Flora*.[2] Thus, when David Don died on 8 Dec, leaving the chair of botany vacant at King's College, London, Watson could expect Hooker's support of his application. Watson published Hooker's letter as the first of seven in the pamphlet which he submitted as an application for the position.

Watson also obtained letters of support from Robert Graham and two other botanists he had known in Edinburgh, Balfour and Robert Kaye Greville. Then Watson lamented to Hooker that Graham had 'so oddly alluded to my supposed theological opinions, that I cannot use his' (n.d., no. 245). Greville mentioned both Watson's 'methodical turn of mind' and his interest in the philosophy of vegetation, and said the college should congratulate itself if it appointed him. Balfour stated that

---

[1] He wrote to Hooker on 3 April 1837 that he had agreed 'to give the botanical lectures in the Liverpool Medical School this season', and on 10 January 1842 he reminded him: 'I have once delivered the regular medical course of 50 lectures, namely, at the Liverpool School of Medicine.'

[2] Watson 1842b. The suggestions for *British Flora* are in Watson's letter of 5 December 1841.

he had been intimately acquainted with Watson while Watson was a medical student, that Watson had distinguished himself as President of the Royal Medical Society of Edinburgh, and that his botanical knowledge highly qualified him for the chair at King's College, London. The zoologist John Edward Gray, who was President of the Botanical Society of London, wrote of Watson's 'great energy and industry' in support of that Society;[3] William Henry Duncan, MD, Lecturer on Materia Medica in the Liverpool Royal Institution School of Medicine, attested that Watson had delivered the botanical course there in 1837; his friend Thomas Seddon Scholes and the surgeon Daniel Noble spoke well of Watson's ability as a lecturer, revealed in his phrenology lectures at Manchester in 1839.

When he wrote to Hooker on 19 January, Watson understood that the candidate would be chosen by eight professors of medicine, but ten days later he heard that the Bishop of London was chairman of the search committee. He confided to Balfour (29 January 1842): 'I do not know whether I am theologically disqualified or not; but if that be only doubtful, political liberalism & phrenological propensities will not improve the matter.' Hooker seems to have advised sidestepping religious controversy, and Watson assured him (rather incorrectly, n.d., no. 244) that he had never overtly disassociated himself from the Church of England, despite his disbelief of many Church doctrines.[4] While he thought that George Don and himself were the strongest known candidates, he nevertheless suspected that Babington would get the position, because of his acquaintance with one of the professors of medicine and his active support of the Church.

By the end of January, Watson had heard from M.W. Bowman, Demonstrator of Anatomy at King's College, that theological beliefs would not disqualify a candidate (n.d., no. 248). Reassured, in his printed application pamphlet Watson explained his former support of phrenology and his nominal support of the Established Church, explaining about the latter: 'I could not honestly subscribe the Thirty-nine Articles, although I do not take upon myself to deny any one of them.' This statement may not be inflamatory, but it is quite distant from the sentiments of the Professor of Botany at Cambridge, John Stevens Henslow, who 'should be grieved if a single word of the Thirty-nine Articles were altered'.[5] Hooker informed Watson that Professor John Frederic Daniell objected to the statement on religion, to which Watson replied: 'I had no wish to mislead them; and I am quite sure, you would not

---

[3] Gray was President of the Botanical Society of London 1836–57 and Superintendent of the Natural History Department of the British Museum 1840–75. Desmond 1977, 264. Gunther 1974 and 1975. Stearn 1981, passim.

[4] Watson to Hooker, n.d., no. 245 and 246. Watson had announced that he was 'a Dissenter from the Church of England' in his quasi-anonymous autobiographical sketch, Waston 1839c, 269.

[5] Quoted in Darwin 1959a, 64–5.

have wished to recommend a candidate who could have been afterwards objected to for having concealed an unfavorable circumstance' (n.d., no. 250). On 19 February Watson wrote to Professor Daniell to inquire whether a decision had been reached. In his letter he volunteered some observations on the strengths and weaknesses of himself and two other leading candidates – a novel idea which could not have helped his cause.[6] Daniell replied that the committee recommended postponement of an appointment, and that Dr John Forbes Royle (Professor of Materia Medica) would give the botany lectures for 1842.

Babington had told Watson that he had not submitted an application for the position, but that he had written to Dr. Royle 'expressing his willingness to fill the Chair, provided Dr. Royle lectured for him this season'. Watson interpreted the committee's decision as indicating the Chair was being held for Babington, and withdrew his application.[7] Although his interpretation was correct, Babington had informed Balfour (8 February) that King's College could not meet his demands, and that the position would probably go to Watson, who 'is far more likely to become a popular lecturer than I was, as he will, I think, take just as much pains and has several advantages that I do not possess'. The committee attempted to change Babington's mind, but he declined again on 26 April.[8]

Meanwhile, zoologist Edward Forbes was dredging for marine animals in the Mediterranean and Aegean Seas. He wanted a permanent position, and his friends in London pressed his case before the search committee. Later, Watson explained to George Combe (15 February 1843) that the King's College charter requires that professors 'be Church of England men,' and that he had withdrawn his application 'on understanding that phrenology and theology were both against me'. Nevertheless, 'on mere botanical grounds I ought to have been elected before Edward Forbes, who was appointed – after I had withdrawn'.

In 1846 Hooker mentioned that a position was likely to be created for the head of a botanic garden in Cape Town, South Africa, and on 26 April Watson responded that he was interested. Watson had decided he might enjoy it, since 'I am fond of the practical management of a garden, & should no doubt speedily become interested in South African botany generally.' On the other hand, there were practical difficulties: it would cause him property losses in England, and he would have to buy a whole botanical library before leaving. Hooker promptly responded that he would recommend Watson to the Secretary for the Colonies. Watson replied (28 April) that he could not get ready within the three or four months which Hooker thought

---

[6] Watson sent extracts from his letters to Daniell, dated 19 and 24 February, to Babington on 24 February 1842, in whose correspondence they remain.
[7] He explained this in letters to Hooker, 24 February, and Balfour, 25 Februay.
[8] Babington 1897, 111 and 285.

adequate. Watson decided on 4 May, that while the pros and cons were about equal, 'the balance is perhaps finally determined against going, by the wish to complete certain fancies in botanical geography'.[9] The situation at Cape Town was possibly more complex than Watson realized, and his backing out was probably best for him and for Cape Town.[10]

Although Watson was reluctant to leave England permanently under economically disadvantageous conditions, he was interested in going a shorter distance on better terms. In July 1846, he heard that two new colleges would be established in Ireland, and decided to apply for a chair of botany at one of them. While he might accept a chair at either, he wrote to Hooker (8 August 1846) saying that he decidedly preferred Cork over Belfast, 'for the mildness of climate, & because I should like to explore the south & west of Ireland, botanically'. Although the plans for these colleges were still being developed, he decided to compile a more elaborate booklet of letters and documents for his candidacy than before (15 pages in 1842; 54 pages of recommendations and supporting information in 1846). Robert Graham had died the previous year, otherwise he might have had another chance to say something that Watson would publish. Forbes was not asked, and since Babington is absent from the booklet, he might not have been asked either.

The botanists he did solicit included all six professors in the British Isles, even though Balfour was the only one Watson knew personally. He had, however, met and corresponded with George J. Allman, Professor of Botany at the University of Dublin, and he had been in contact with four other Irish botanists who were not professors.[11] British professors whom he did not know were requested to comment on his publications. Walker Arnott, who had become Professor of Botany at

---

[9] His 4 May letter to W. Hooker must not have been his final decision, however, because he wrote to J. Hooker on 15 July [1846]: 'Were I assured of removing to the Cape, I might have suggested that W. Hawitson should look whether my cottage would meet his wants.' Watson sent his ideas for such a garden on 4 May 46 and again on 14 March 1847.

[10] There were already two private gardens of note there, the one owned by the German Carl F.H. Ludwig being more prominent than the other, owned by the Englishman John Ross. Although the plans of Hooker and the Secretary of the Colonies are unknown to me, the need for diplomacy in that situation probably exceeded Watson's capacity. Bradlow 1965, especially 67–8. The following negative information was kindly furnished by Sylvia M.D. FitzGerald, Chief Librarian and Archivist, Royal Botanic Gardens, Kew: 'As far as I can tell from our indexes, Sir William Hooker did not correspond with John Ross of Cape Town; we have letters from C.F.H. Ludwig to Hooker dated between 1835 and 1844, but nothing for 1846.'

[11] Allman wrote a favorable letter, and John Ball, William Henry Harvey and Dr Thomas Taylor signed petitions of support, in *Testimonials in Favour of Hewett Cottrell Watson, as a Candidate for a Chair of Botany in Ireland* (Watson 1846a), 37, 7, 11 and 15. Watson had also corresponded with William Andrews, but did not request his support either because Andrews himself was applying for a chair or because Andrews's veracity was questionable. Watson to Combe (8 September 1846). On these botanists, see Desmond 1977 and Praeger 1949.

Glasgow in 1845 when Balfour won the Edinburgh chair, wrote a letter which Watson disliked but nevertheless published:

> Hewett Cottrell Watson, Esq. is well known to me as the author of several works on the geographical distribution of plants, particularly in connexion with the British Isles; and also of several detached papers on species in various periodicals. I have no doubt, from his talents and zeal, that he will prove as efficient a teacher of botany as most others who at present lecture on that science.[12]

The last statement is a compliment if one assumes that most of the others were good lecturers, and this is probably what was intended. However, to a veteran of verbal combat in the trenches of phrenology, the ambiguity seemed insulting.

Watson characteristically over-reacted to this testimonial in a letter to Balfour (14 August 1846): 'Dr. Arnott, indirectly, gives you, me and others a sneer in the testimonial sent to me. He looks with contempt upon all who are not devoting themselves to the technicalities of classification.' If Watson really believed this, he could simply have omitted Walker Arnott's statement from his printed testimonials, as he had omitted Graham's in 1842. Since he included it, he was merely expressing to Balfour his resentment over a neutral evaluation of his achievements. In 1845 Walker Arnott had corrected in print some of Balfour's identifications of Brazilian orchids,[13] and Watson may have imagined that this embarrassment would make Balfour sympathetic to his complaints. Watson now conceded that Walker Arnott had at least a limited competency while simultaneously attacking his narrowness: 'Excellent as he is in this [ability to classify], I pity his students; for they will care little about it, and it must be thoroughly unpalatable to most beginners.' Watson obtained a brief testimonial from the amateur botanist and phrenologist Sir Walter Calverley Trevelyan, to whom Watson sent a note written on a copy of the testimonial booklet: 'I trust that you will see no reason to regret the favour of your own recommendation so kindly given to me.'[14] He even wrote to his childhood mentor, Dr Edward Stanley, now Bishop of Norwich and President of the Linnean Society of London, whose nephew might carry some influence.[15]

---

[12] In Watson 1846a, 36.

[13] Arnott 1845.

[14] In Watson 1846a, 28: 'I believe that Hewett C. Watson, Esq. is well qualified to hold a Botanical Professorship, and that he is competent to discharge its duties with credit to himself, and with profit to his pupils.' The copy of *Testimonials* Watson sent to Trevelyan with note dated 27 October 1846 is now in the British Library. On Trevelyan as a botanist, see Desmond 1977, 618. On Trevelyan and phrenology, see Gibbon 1878, I, 136, and II, 119; Trevelyan 1978, 14 and 19. Watson's letters to Trevelyan soliciting his recommendation (5 August 1846) and thanking him for it (12 August 1846) are also extant.

[15] Mentioned in a letter to Combe, 15 August 1846.

Support from botanists was only half of what Watson needed. He was not so fervently committed to democratic ideals that he spurned aid from influential connections. Testimonials in his booklet also came from his brothers-in-law, William Gorton, John Bulkeley Johnson, John Horatio Lloyd, James Spence and Henry Stewart, and the two Wakefields who signed must have been relatives of his late brother-in-law, Gilbert Wakefield (d. 17 January 1846). Watson stated (p. 19) that his position in society was indicated by the names of these and other listed relatives. Sir William Hooker volunteered to speak on his behalf to the Anglican Archbishop of Dublin, Richard Whateley, for which favor Watson responded with appreciation (2 September 1846):

> Mr. Combe, who is on terms of friendly correspondence with His Grace, and the Rev. L[ambert] W[atson] Hepenstal, a cousin of mine, who is on visiting terms, have both written to him, & one or other will doubtless inform me when he arrives at home. I believe that my brother-in-law, the Rev. Dr. Stewart, uncle of the Irish botanist of that name, will also communicate with the Archbishop through a friend, tho' not personally acquainted. But since none of these gentlemen are botanists, your recommendation of me ought to have more weight.

Later on, Watson heard that the Archbishop would not offer advice unless the government asked for it, and therefore Watson wrote to Hooker (8 September) that he need not bother to contact the Archbishop for him after all. Watson arranged for a prominent Irish member of the House of Commons (probably Mr. Wyse, whom he mentioned to Combe, 5 August) to deliver his testimonials to the man who was presumably the most influential officer in the decisions, Henry Labouchere, Chief Secretary of State for Ireland.

His dealings with King's College had taught Watson to remain silent on religion. Testimonials from churchmen could indicate that he would not embarrass either church or government. Phrenology, however, was a different matter. Not only had he increased the number of recommendations from phrenologists since the 1842 booklet, he also listed his most important phrenological writings after his botanical ones and defiantly explained:

> Some persons may deem this second class of writings to be no recommendation for a Chair of Science. My own opinion is the contrary. I have no hesitation in saying, from observation and some experience, that a higher and a wider range of thought is required for writing on Phrenological subjects than is needed for writing with equal success on Botanical subjects. (p. 50)

While his phrenological friends would have been pleased, these words more likely hindered than helped his cause. However, Archbishop Whately would read this statement sympathetically, and Combe encouraged Watson to go see him (29 September 1846):

> ... although not officially concerned, [he] could certainly give you a lift forward if moved by a proper stimulus. The key to his good will (I regret to say it) is through his love of Approbation. If you will carry an introduction to him from his brother bishop, and throw out some natural praise of his Logic, Rhetoric, and Political Economy, or other books, and listen to his talk, you will make an impression that will, or may, result in good ... He will welcome you as a phrenologist.[16]

Despite much encouragement, Watson vacillated. Ambivalence, he explained to Combe (4 October 1846), was partly a hedge against disappointment and partly a lack of conviction that he really wanted the position. Writing to Combe again on 22 October, he virtually elevated ambivalence to a policy:

> ... the botanists think that Phrenology has no business to be brought into the recommendations; and other friends pronounce me very injudicious, in not rather avoiding it. I tell them, that if a botanical chair in Ireland had been better worth having, I might have the more readily consented to act like 'a man of the world', in striving for it.

Combe suffered from no such ambivalence and pressed Watson's case before the Archbishop of Dublin who came to Edinburgh for a visit. Combe saw him several times and informed Watson (6 November) that:

> ...he is extensively and accurately acquainted with the details of Botany, and particularly with the physiology of plants. This is the department which engages his interest. When he mentioned this, I told him that this was your peculiar department, that your works in it are highly esteemed, and asked if he had seen them. He said that he had not. I took occasion to say as much about you and in your favour as a candidate for an Irish Professorship as I could with propriety; and I leave you to judge whether in these circumstances you should not send him your testimonials, letter of introduction, and a copy of any of your writings that you may think calculated to interest him.

Watson clearly had substantial, well-connected support.

---

[16] Rigg 1899 includes an extensive list of publications in his biographical account of Richard Whately.

The government moved slowly in establishing these colleges because of a difficult political issue: whether they would be Catholic or secular.[17] Watson commented pessimistically to Combe (10 November):

> There is neither trust in the Government nor trust in the Irish; and the very want of this trust will go far towards realizing the unsuccess.

Since both Combe and Bishop Stanley prepared the way for him, Watson wrote to the Archbishop. When he received no response, he indulged in typical hostile speculations as to why not (in letters to Combe: 29 December 1846 and 13 January 1847). Whately finally wrote to Combe, if not to Watson, and Watson was satisfied with the letter which Combe sent on to him (he returned it on 2 February 1847).

With all his to-ing and fro-ing, Watson never seems to have had as strong a desire for an Irish chair as he had had for the one at King's College London. Combe, with Watson's interests at heart, threw cold water on the Irish possibility in a letter (27 September 1848) written from Edinburgh just after returning from a trip to Dublin and Belfast:

> I was introduced to Sir Robert Kane, the Principal of [Cork] College, and regret to say that my estimate of him was not such as to lead me to make any advances to him. He is a self-raised man (all in his favour), a Roman Catholic (nothing against him), has a large anterior lobe, speaks *ore rotundo* in very measured sentences, and is said to be an able chemist and to have written a good book. What then do I dislike? He gave me the impression of a man who had no greatness of views, no depth of principle; but of one who had risen on the Roman Catholic interest into his present situation, who would use his [influence] in favour of that interest, and in whose estimation the great interests of humanity would stand second to the party and religious politics of Ireland. In short, I inferred that he would fill his college with Roman Catholics, and that a man like you would be an aversion to him ...

After reading this evaluation of the President of Queen's College, Cork, Watson lost interest in a professorship there. His decision, as explained to Combe (12 December 1848), was not based solely on Combe's estimate of Kane. In addition:

> Botany and Zoology are united as one Chair, and not, as I think, as should have been done, Botany and Agriculture. I should conceive it far more interesting to teach a science and its applications, than to teach two sciences unapplied to

---

[17] Gwynn 1948.

anything. Two and a half years have been added to my age; and that time has not rendered Ireland more desirable as a place of abode, but very much the reverse. Moreover, £200 a year salary, for a shorter number of years, is much less remunerative for the cost and trouble of a distant removal, than (as first said) £300 a year for a longer term.

Concerning salary differentials, Watson commented:

It is worthy of note, how much the ancient idea predominates even in the new Irish Colleges. The Teacher of Greek, and the Teacher of Latin, are to stand highest in Salary – £250. Yet they require no apparatus, no collection of objects, and not more books than the Teachers of other things.

Did Combe judge Kane too harshly and possibly thwart Watson's career unnecessarily? Despite Combe's attempt to be fair, there is a possibility of a bias by a Scotsman against an Irish Catholic. He may have been correct in believing that Kane's religious and political loyalties were considered when he was made President. This may have been part of a compromise, to establish secular colleges while letting Catholics have a prominent role in running them. Combe overestimated either Kane's determination to hire only Catholics or his ability to do so. Kane's biographer says he earned a high reputation as an administrator.[18] Nevertheless, one episode during Kane's administration provides an insight into how Watson and Kane might have interacted. The Professor of Anatomy, Benjamin Alcock, was a well-qualified surgeon and physician. In 1853 he claimed that the administration expected him to violate the Anatomy Act, which regulated the procurement of cadavers for dissection. Not enough documents survive for the historian of this episode to make a definitive assessment of blame between Alcock and Kane, but Alcock had to resign his position.[19] Even if we shift most of the blame to Alcock and presume he was combative and diplomatically inept, one sees the same qualities in Watson. Combe may have been wise in advising Watson not to continue his application.

When the new colleges finally hired professors of natural history in 1849, Watson wrote the announcement for *The Phytologist*. He used the occasion to inform fellow botanists that he had withdrawn, and had not been passed over. By then there were three new Queen's Colleges in Ireland instead of two, and Watson wrote enthusiastically about the appointment of George Dickie (1812–82) at Belfast:

---

[18] Reilly 1955, 406. On his previous career as a chemist, see also Kelham 1973 and Wheeler 1945.
[19] Alcock apparently emigrated to America; O'Rahilly 1948.

the papers which he has already published on botanical subjects, being such as to place him in an elevated and honorable position among scientific naturalists; and giving promise, we trust, of such valuable exertion yet to be made by him for the promotion of science.[20]

Watson had good reason for this warm endorsement: Dickie was practically the only botanist who had followed Watson's lead in making careful studies on the altitudinal influences upon the distributions of plants in the Scottish Highlands.[21] While in Belfast, he collected material for his *Flora of Ulster* (1864), though in 1860 he returned to his native Aberdeen as Professor of botany in its university.

---

[20] [Watson] 1849k. The other two appointments were Rev. William Hincks (1794–1871) at Cork (his home town), who was to teach there 1849–53 and Alexander Gordon Melville (1819–1901) at Galway. On Dickie and Hincks, see Desmond 1977, 183 and 310; Praeger 1949, 71 and 102. Michael Mitchell tells me that Melville taught at Galway until 1882. Andrew Smith Melville, mentioned in Desmond 1977, 434, was A.G. Melville's son. Dickie is listed in Stafleu & Cowan 1976–92, while Hincks and Melville are not; see I, 643–4. Dickie, Hincks and Melville are not among the Irish botanists discussed in Nelson 1993.

[21] He praised Watson's work in Dickie 1843a, 133; see also Dickie 1843b and Dickie 1847.

Chapter 9

Professional Relationships with Forbes, Babington and Balfour, 1833–59

I think that we shall have an attack upon Forbes from the enemy soon.
Babington to Balfour, 26 October 1846

Watson maintained friendly relationships with George Combe for thirty years, and with William Hooker for twenty-two years, for they served as mentors to him in phrenology and botany. It was much more difficult for Watson to remain cordial with contemporaries whom he could view as rivals. This difficulty is illustrated by his relationships with Edward Forbes, Charles Cardale Babington and John Hutton Balfour, all a few years younger than him. All three held academic positions, and their relationships with Watson were thus professional ones.[1] They were friends with each other, but their relationships with Watson began warmly and gradually cooled.

Conflict with Edward Forbes

Watson had not competed for the chair at King's College, London, with the candidate eventually hired, Edward Forbes, since Watson withdrew his name from consideration while the position was being held for Babington. Forbes was hired only after Babington twice refused it. Nevertheless, when the dust settled, Forbes, a zoologist, held a chair in botany which Watson had wanted.

Forbes (1815–54) was from a comfortable middle-class family on the Isle of Man. He turned to natural history as a diversion from an adolescent illness. He entered the University of Edinburgh in November 1831, shortly before Watson left it. Like Watson, he took the medical curriculum without completing the degree, because he lacked a desire to practice medicine.

---

[1] These relationships were professional ones, even though I accept David Allen's observation that Watson was only a proto-professional – a full-time botanist who nevertheless failed to obtain professional employment. Allen 1985.

While both men expected to live, at least in part, on inherited wealth, Forbes's expectation ended in 1841, because of his father's financial reverses. At the time, Forbes was dredging for marine animals in the Mediterranean and Aegean Seas. His close friend and fellow marine zoologist, John Goodsir, persuaded the search committee for the King's College chair to delay their appointment until Forbes returned for an interview in October 1842. The committee decided in his favor before he reached London. Forbes regretted that the chair was not in natural history, 'for he felt diffident of his botanical acquirements, and even alleged that nothing but the absolute necessity of obtaining some remunerative employment would induce him to accept a lectureship of botany'.[2] Yet he was not lacking in botanical interests and knowledge. He had studied botany under Robert Graham, and he had already published several brief papers on botany. Watson himself had published Forbes's list of flowering plants and ferns on the Isle of Man in *The New Botanist's Guide to the Localities of the Rarer Plants of Britain* (1837). Forbes was also a charter member of the Botanical Society of Edinburgh.[3] He was certainly competent to teach botany to medical students, which must have been the main concern of the college.[4]

Forbes decided that the insights which he had gained on the zonal distribution of marine animals might apply also to the distribution of terrestrial plants. Being trained in geology, he understood that paleontology could add to the knowledge of the distribution of species. Surveying the distributional patterns of British plants, he concluded that there were five distinct divisions of the vegetation, each of which was tied to a stage of the geological history of Britain. His line of reasoning is what one might expect from an evolutionist, which he was not, and paradoxically, Watson, who was an evolutionist, avoided speculating along these lines because he lacked training in geology.[5] At the meeting of the British Association in Cambridge, June 1845, Forbes announced his new interpretation: 'On the Distribution of Endemic Plants, More Especially Those of the British Islands, Considered with Regard to Geological Changes'. Forbes was introducing a genuine novelty in his effort to tie the modern vegetation to the geological history of Britain. He saw his contribution as being a theoretical interpretation based largely upon data already published. His presentation was well received.

---

[2] Wilson & Geikie 1861, 307. See also: Browne 1981, Egerton 1972 and Mills 1984.
[3] Desmond 1977, 227–8. Fletcher & Brown 1970, 113. Rehbock 1979, 181; Rehbock 1983, 127–8. Stafleu & Cowan 1976–92, I, 852.
[4] The popularity of his lectures with medical students and others is attested by Wilson & Geikie 1861, 345.
[5] By 1848 Forbes began developing his theory of polarity, which was his alternative to ideas of either evolution or progressionism. Browne 1983, 149–55 and 171–2; Mills 1984, 377–9; Rehbock 1983, 102–13 *et passim*.

Watson was not present, but an abstract of the lecture, which he judged (correctly) to have been written by Forbes, appeared not only in the *Report of the British Association*, but also in the *Annals and Magazine of Natural History*, the *Athenaeum*, the *Literary Gazette*, the *Gardeners' Chronicle*, and the *Phytologist*.[6] The abstract did not specify the sources of information on which Forbes had based his conclusions. Watson suspected that Forbes's crucial source was his own *Remarks on the Geographical Distribution of British Plants; Chiefly in Connection with Latitude, Elevation, and Climate* (1835) – which employed six regional subdivisions, not five. He checked and found that Forbes had borrowed a copy of the *Remarks* from the library of the Linnean Society of London on 16 June, about one week before he read his paper.[7] While such acknowledgment would have been desirable in the *Report of the British Association*, it was not an absolute requirement for a summary. That Forbes was probably insufficiently sensitive about such matters is indicated by a remark which his friend, Darwin, made to Joseph Hooker some six months later:

> I think Forbes ought to allude a little to Lyell's work on nearly the very same subject as his [recent] speculations; not that I mean that Forbes wishes to take the smallest credit from him or any man alive: no man, as far as I can see, likes so much to give credit to others, or more soars above the petty craving for self-celebrity.[8]

Forbes learned of Watson's anger and attempted to satisfy him the next year when publishing a fuller exposition of his ideas: 'On the Connexion between the Distribution of the Existing Fauna and Flora of the British Isles and the Geological Changes Which Have Affected Their Area'. He was also standing on firmer ground this time by including animals – subjects of his own research – as well as the plants discussed the previous year. His introductory remarks stated both what others had contributed and what he wished to achieve:

> Having assumed the doctrine of specific centres [of creation] as true, the problem to be solved is, *the origin of the assemblages of the animals and plants now inhabiting the British Islands*. The zoological works of Fleming, Jenyns, Yarrell, Bell, and W. Thompson, have enumerated the species and treated of the distribution of our indigenous animals, those of Smith, Hooker, Lindley, Babington, Henslow, and especially Watson, have done the same service for our *native flora*; but the history of the *formation*, if I may so say, of that fauna and

---

[6] Forbes 1845; see Bibliography for citations of the summaries.
[7] Watson 1847–59, I, 468 and 472.
[8] Darwin to Hooker, 25 February 1846; quoted in Darwin 1985–, III, 294.

Figure 11 Edward Forbes holding a plant (1849), by T.H. Maguire.

flora, remains to be investigated. This essay is offered as a contribution towards such a history.

He explained the similarities and differences between his own fivefold and Watson's sixfold division of the British vegetation, and in hopes of avoiding conflict with Watson, Forbes stated:

> ... such inquiries as the present ... could not be entered upon with advantage unless there were abundant data gathered together. The essays of Mr. H.C. Watson may be cited as among the most remarkable, and to them I refer geologists who would wish to learn more respecting our indigenous flora, than it is here necessary to state.[9]

Forbes published his study before the 1846 meeting of BAAS, but Joseph Hooker, who did not plan to attend that meeting, doubted that it would appease Watson. He lamented to Darwin:

> This probable fracas between the 2 Geographers distresses me, for they are almost the only 2 men who have looked on British Flora with the eyes of philosophers. Watson in particular ranks in my opinion at the very head of English Botanists, whether for knowledge of species or their distribution; he first wrote philosophically upon them & his works are of the highest order.[10]

Watson was not one to be silenced by flattery accompanying what he believed was an unwarranted revision of his work. He so informed Forbes when they met in September at the BAAS meeting in Southampton, where Forbes gave him a copy of his study. Forbes also presented a copy to Hooker to show him that he had adequately acknowledged his use of Watson's work.

However, the acknowledgment also seemed inadequate to Hooker after Watson pointed out some of Forbes's 'outrageous' blunders (16 September). Disappointedly, Hooker confided to Darwin (28 September):

> I have not seen Forbes since studying his paper & really do not know what to say when I do, for he will be sure to ask me about it, & most unfortunately he does not seem to know the Geographic Distrib. of the English Plants. I must confess to have taken his modification of Watson's types of vegetation as correct, & this for granted, but I had occasion to look closely at them the other day & find his

---

[9] Forbes 1846a, 336 and 342.
[10] Hooker to Darwin, before 3 September 1846; quoted in Darwin 1985–, III, 336–7.

S.E. Flora, numbered III., to be altogether a fallacy: all or almost all the 20 species on whose supposed presence he founds it, being as common or more in the W. or N as in the E. or S. & some of them not existing in the S.E. at all! or if so as introduced species. I now see the cause for Watson's being so peculiarly savage & offering me proof that all that is correct is mere plagiarism.

Hooker still felt convinced that Forbes was innocent of 'intentional piracy', but was pessimistic about convincing Watson not to denounce Forbes in print. To enable Darwin to judge the situation for himself, Hooker promised to send him one of Watson's works.[11] Darwin, who had attended the Southampton meeting, wrote to Hooker about the Watson–Forbes encounter (2 October):

> I missed having a look at H. Watson. I suppose you heard that he met Forbes and told him he had a severe article in the Press. I understood that Forbes explained to him that he had no cause to complain, but as the article was printed, he would not withdraw it, but offered it to Forbes for him to append notes to it, which Forbes naturally declined ...[12]

Watson contended that Forbes lacked the capacity to judge and reject Watson's sixfold division of British vegetation, and that Forbes's belated acknowledgment did not undo the fraud that was still being perpetrated. Watson frequently corresponded with Babington about the accuracy of available information on the distribution of plant species in England, Scotland, and Wales. Within this context, he explained to him how Forbes had incautiously compiled the data for his 'Connexion' paper, while still leaving vague his precise debts to others (26 October):

> He turned over the pages of your Manual, & trusted to the E[ngland], I[reland], S[cotland], O[rigin unknown]; & ignorant himself on the subject he was writing about, has misunderstood & misapplied your letters & notes of localities most egregiously. Of course the misprint of the E to *Erigeron alpinus* is faithfully adopted, & the locality boldly fixed in Wales. He ought at least to have acknowledged how *specially* he was indebted to that novel feature in your Manual, instead of putting forth his lists as though they were the results of his own individual knowledge & investigations.

---

[11] Hooker to Darwin, 28 September 1846; in Darwin 1985– , III, 342.
[12] Darwin 1985–, III, 346. Darwin's reading of Forbes is discussed by Browne 1978, 353–4; 1983, 126–7.

I need not caution you not to commit yourself to the utterly insufficient evidence, full of errors, which he has adduced to support views announced to the world *before* he had troubled himself by any inquiry for proofs of those views.

If Watson had been outgoing and personable and Forbes withdrawn and caustic, then surely Forbes's reputation among botanists, if not British scientists generally, would have been dead by this time. Instead, Watson was withdrawn and Forbes was outgoing. Forbes's character and competency had already been evaluated by British naturalists, who judged him to be honest and competent, if incautious. Forbes sent two letters defending his behavior to Joseph Hooker,[13] and he undoubtedly conveyed comparable messages to other botanists.

Babington had written a recommendation for Forbes in the latter's unsuccessful quest for the Chair of Natural History at the University of Aberdeen in 1841,[14] and his previous judgment was not demolished by Watson's damaging revelation. Babington's response to Watson's attacks on Forbes was sent the same day (26 October) to Balfour:

I think that we shall have an attack upon Forbes from the enemy soon. H.C.W. writes to me in such a way as to make it almost certain. I fear that Forbes has taken my E. S. I. as his guide. Those must in many cases be incorrect, and at all events only pretend to be an approximation of the truth.[15]

Balfour also had defended their friend in August and September by taking some students on two field trips – one north to Clova, Glen Isla and Braemar, and the other south to the Isle of Wight – to illustrate Forbes's Scandinavian and Devon Floras, respectively. He later reported their findings orally to the Botanical Society of Edinburgh, and news of those reports was published in *The Phytologist*. Now Babington's bad news caused Balfour to alter his account of the findings before publication; it reached print in the form of a report on field trips, not as a verification for Forbes.[16]

Watson's account of Forbes's sins and the inadequacy of his restitution appeared in an appendix to *Cybele Britannica* (vol. 1, 1847, pp. 465–72). He sent a copy of the volume to Babington, who responded with general praise, but also with regret

---

[13] Forbes to J. Hooker, 31 October 1846 & 22 April 1847; quoted in Rehbock 1983, 180–1.
[14] Reproduced in Babington 1897, 98.
[15] Babington to Balfour, 26 October 1846, reproduced in Babington 1897, 299.
[16] For brief accounts of those trips, see *The Phytologist*, 2 (1847) 740–1 and 862. Balfour's account on the 3-week excursion to Braemar, Glen Isla, Clova and Ben Lawers, published in the *Edinburgh New Philosophical J.* (July 1848), is partly quoted in Fletcher & Brown 1970, 133–4.

that Watson had not left some things unsaid. Defensively, Watson replied (15 April 1847): 'I consider [Forbes's] behaviour to me fully warranted the strictures. Private information (of an oral admission, made by F. to an acquaintance) would make the case much stronger against him, but cannot be publicly used.' We may assume that Watson had condemned Forbes to the same category of The Damned to which he had earlier consigned his father and brothers. Forbes's only published reaction to Watson's *Cybele* complaint seems to have been a wry comment in a review of some botanical books (1852), in which he characterized Watson as 'indefatigable and deservedly illustrious in statistics, but grown misanthropic by working overmuch when in ill humour, [who takes] a melancholy pleasure in attributing evil motives to his fellow-labourers'.[17]

Watson's attack on Forbes harmed both men. He was fully justified in defending his own scholarship and exposing sloppy aspects of Forbes's. However, after Forbes tried to make amends in 1846, Watson's unwillingness to moderate his complaint in 1847 resulted in the scientific community judging him as being too harsh.

Botanical Colleagues: Watson versus Babington and Balfour

Both Babington and Balfour were affable, diligent men who had successful university careers. Born in the same year, 1808, they came from professional families: Babington's father was a physician, clergyman and amateur botanist, while Balfour's father was an army surgeon and later a printer and publisher in Edinburgh. Both botanists were devout Christians, and they became lifelong friends.[18]

Babington had an early interest in entomology but was won over to botany at Cambridge University by Professor Henslow. Unfortunately, as Babington once explained to Balfour, the Colleges were rich, but the university was not. Henslow could not live on his professorship and therefore became an ordained priest in the Church of England. He went on to serve as a parish priest, first in Cambridge and later in Hitcham. For two years Henslow was an absentee clergyman who at times visited his parish, but in 1839 he moved to Hitcham and became an absentee professor who returned to Cambridge for five weeks a year to lecture.[19] For over two decades Babington was the poorly paid resident Cambridge botanist. His fidelity

---

[17] Forbes 1852, 212. See also Rehbock 1983, 178.
[18] On Babington (1808–95), see Allen 1999, Babington 1897, Desmond 1970a, Desmond 1977, 25, Walters 1981, 67–70. On Balfour (1808–84), see Balfour 1913, Desmond 1970c and 1977, 32–3, Fletcher & Brown 1970, chaps 10–12. On both, see Freeman 1980, nos. 133–139 and 178–190; Stafleu & Cowan 1976–92, I, 83–5, 112.
[19] Barlow 1967. Desmond 1977, 303. Russell-Gebbett 1977.

was finally rewarded in 1861, after Henslow died: Babington was elected to the Chair. Perhaps financial constraints inhibited him from marrying earlier than he did – in 1866, at age 58.

Babington never did much teaching, but he was active in a number of organizations. He was a founder of the entomological and the antiquarian societies in Cambridge, and assumed important roles in the Ray Society and in the Botanical Society of Edinburgh. He published local floras for Bath (1834), the Channel Islands (1839) and Cambridgeshire (1860), and he also wrote a synopsis (1846) and a monograph (1869) on the British Rubi – attempting to disentangle the blackberries and raspberries. Although W.J. Hooker published successive editions of his *British Flora*, Babington decided there was room for a rival treatise and began publishing editions of his *Manual of British Botany* in 1843. Hooker had many irons in the fire – mostly unrelated to the British flora (as Watson was wont to remind him). In order to keep up with the competition, Hooker brought in George Walker Arnott to help him update his sixth and seventh editions. Babington, on the other hand, worked mostly on the British flora, and published eight editions of him *Manual* without a co-author.

Balfour was drawn to botany by the lectures in Edinburgh of Professor Graham. He obtained his MD degree in 1832 and continued his medical training in Paris for two years. He then returned to Edinburgh to practice, but still found time to organize and run the Botanical Society of Edinburgh (founded 8 February 1836, with Graham as *nominal* president). In 1840 he began teaching classes in botany for the Medical School, and in 1841 he succeeded Hooker at Glasgow. That provided the experience needed to win the competition for the vacant chair at Edinburgh in 1845. Balfour became too deeply involved with teaching, the Botanical Society, the Botanic Garden, and the administration of University and Medical School to establish a strong record of publications in botany. He did publish textbooks (1849, 1852 and 1854) and books relating botany to religion (1851, 1857 and 1870), and he co-authored a *Flora of Edinburgh* (1863), but he published only about fifty brief botanical articles, compared to Babington's 132.

Watson had known Balfour at Edinburgh, where they may have taken classes together, and they certainly went on Graham's field trips together. Much later, Watson wrote to him (8 February 1871) that he had left Edinburgh in 1833 'with feelings of the utmost cordiality towards yourself, Brand, Campbell & others'. This recollection is borne out by their correspondence, which began by March 1836. By June Watson was interested in joining the new Botanical Society of Edinburgh (BSE) and participating in its program for exchanging plant specimens (which had been Watson's idea in the first place[20]). Watson responded to Balfour's news that the

---

[20] Allen 1965.

phrenologist Browne had married his sister by trying to recruit Balfour for the movement ('The science [of phrenology] has run its ordeal almost completely, & is on the eve not merely of general reception, but also of becoming the groundwork of vast changes in society, political & otherwise.') How and when Watson and Babington met is unclear, but it seems to have happened because of their mutual publications. Watson had freely distributed copies of his *Outlines of the Geographical Distribution of British Plants* (1832) and had sent one to Henslow. This book then probably inspired Babington to send Watson a copy of his *Flora Bathonensis*. Watson responded (18 April 1834) with warm thanks and praise, and offered to send him a copy of his own *Outlines*, but confessed that it had been too hastily published.

Watson's relationships with these two slightly younger colleagues never ruptured over any cardinal sin the way his relationship with Forbes had. Nor was their well-known piety and his equally well-known skepticism a conscious impediment to their professional involvements, though it might have prevented them from ever accepting his invitations to visit him at Thames Ditton. Although the professional goals of these three botanists were not identical, there was substantial overlap. They all wanted to describe in some precise way the British flora and its distributions. It was a communal enterprise with many participants, among whom they were the leaders.

Babington and Balfour could lead this endeavor in relative harmony, and even where there was direct competition, as between William Hooker's and Babington's alternative guides to the British Flora, no overt friction emerged. But adding Watson to the group produced disharmony. As Babington expressed it to Balfour (23 January 1846):

> I do not think that [Watson] wants to quarrel with either of us but he has a manner which makes it very difficult to keep on good terms with him. He seems to think that he may say and print whatever he likes of others & that they must not even hint at anything on the other side.

This was a valid characterization: Watson would seize the high ground and pretend to make 'objective' judgments about others.

The correspondence between these three botanists shows that scientific judgments about particular plants or groups of plants was not a basis for friction. Watson never seems to have tried directly to convince either of them to accept his evolutionary ideas, and all differences of judgment seem to have been over the status of particular specimens or groups of specimens, with the issue being whether particular variations were important enough to indicate different species or merely varieties within a species. An early example is in Watson's letter to Babington on 15 February 1836:

As to your *Erica Mackaiana*, I really cannot join you in calling it 'n[ew] s[pecies].' But it is highly interesting to me as forming one of the links whereby *E. ciliaris* is joined to *E. Tetralix*. My herbarium now shows a series of specimens by which these two are joined in the shape & distribution of flowers as well as of leaves.

A later example (29 October 1841) hints at theoretical considerations:

I have latterly been looking out for any variations from the typical characters of species, in the belief that we should much better understand plants, by bringing together series of specimens of each species, to illustrate the extent to which each one may vary. – For example, *Linaria repens & vulgaris*, in their typical forms, are very little alike, save in general habit; yet I have specimens running so much from one to the other, as to render it dubious to which species some of the forms should be assigned – You will find examples of these in the parcel [being sent to you].

In the same letter, he both praised and damned the checklist of British plants which Babington had compiled for BSE:

I am delighted with the new edition of the Edinb. Catalogue. It is just what British botanists required, and will be eminently useful. Tho' I am too much a 'lumper of species' to agree in all your divisions, yet, with several it seems likely you are in the right view. I should have preferred a systematic, in place of alphabetic catalogue.

Some people seem to be more impressed with similarities between items, and others more with differences. These inclinations might at times be linked with theoretical perceptions, but are not inevitably so, since the arguments between 'lumpers' and 'splitters' continued in systematic biology after all participants in the debate were evolutionists. Watson may have hoped that his gentle prodding in letters and his published judgments would convert Babington to his own perspective. That never happened, and he complained to Balfour (20 November 1846):

I'm thinking of proposing a new 'rank' in botany to receive plants of the debatable ground, intermediate between species & varieties. And as our friend of the *Manual* appears particularly fond of putting 'Bab' after the names of such debatables, I propose to give these 3 letters a termination corresponding with the other 2 ranks, and so get a good & applicable name for the intermediate ranks. The 3 degrees will stand thus:
   Species –
      Babies –
         Varieties –

... We can say of a Linnean species that it is divisible into so many Babingtonian babies.

Balfour probably defended his friend, but Watson was not deterred. In his last letter to Balfour for many years (30 December 1846) he continued to express frustration with Babington:

He is really a politic man! He has the knack of making other botanists his 'ministering slaves'. But as he is capable in botanical matters, & careful, too – always excepting his ridiculous monomania for species making & name-changing – it is well he should guide others.

Even if Balfour responded to this letter, it is unlikely that he responded to this particular complaint/compliment. There are no further letters for almost eighteen years.

Babington perhaps met Forbes at meetings of the British Association, or maybe at the Linnean Society of London. Their friendship was not so close as to cause Babington to cease corresponding with Watson over the Forbes affair. On the other hand, exchanges between Watson and Babington do not indicate a friendship. Babington referred to Watson (half-jokingly, perhaps) in a letter to Balfour (26 October 1846) as 'the enemy', and yet he persisted in a professional relationship with Watson for two reasons. First, Watson was an active botanist whose judgments were based upon demonstrable evidence which could not easily be dismissed. Second, Babington was bringing out successive editions of his *Manual of British Botany*, and although he was far from accepting all of Watson's judgments, he liked to consider them when making revisions for a new edition. That Babington accepted enough of Watson's judgments to win the latter's approval was attested in his anonymous review of the second edition of Babington's *Manual*, which appeared in *The Phytologist*: 'The former edition of this work was good – very good ... this edition is better.'[21] Watson and Babington were engaged in a common endeavor which transcended personal and theoretical differences. Both knew the other was fully committed to a precise account of the British flora, and each took account of the other's judgments in forming his own. This esteem for Babington's *Manual* probably continued through the later editions as well (the eighth edition appeared in 1881, the year of Watson's death). Watson even preferred the third edition to Hooker's *British Flora*; his candid assessments appeared in 1852:

No doubt the *British Flora* of Sir W.J. Hooker was a very good work originally; and its publication sufficiently opportune and beneficial a score of years ago. But

---

[21] Watson 1847b, 843.

British botany has progressed much during the past twenty years, while the Author of the *British Flora* was directing his own attention almost exclusively to the plants of other and distant regions. Hence, the *British Flora* was falling more and more into arrear, in each successive edition, and was thus making an opening for the successful competitor which appeared in [Babington's] *Manual*. It is left behind and will now never overtake that competitor ...[22]

Hooker had attempted to remedy the situation by bring in George A. Walker Arnott as co-author of the sixth edition, but Watson viewed the remedy as inadequate because he, too, was distracted by exotic botany. To illustrate the difference between these works, Watson pointed out that Hooker and Arnott (1850) were still citing Watson's *Remarks on the Geographical Distribution of British Plants* (1835), whereas Babington (1851) was using Watson's *Cybele Britannica*, two volumes of which had appeared before Hooker and Arnott went to press. Although Babington was undoubtedly gratified to see his *Manual* praised so highly, he may have wondered why Watson had not reproved privately such an obliging colleague as Hooker had been to Watson for more than two decades. Babington's own correspondence with Watson lasted for about another decade.

Botanical Societies: Edinburgh versus London

Balfour organized BSE in February 1836 – about six months before the Botanical Society of London (BSL) was organized. London had long been an active center for botany, but two factors probably inhibited any earlier botanical society there: (1) a variety of competing organizations – such as the Royal Society and the Linnean Society – and (2) no strong university from which to draw in faculty and students. The University of London was only established in 1836.[23] Watson lived at Thames Ditton for some two-and-a-half years before BSL, but he did not have close ties to the amateur botanists who founded it. On 31 March 1836 (which is the earliest known evidence that it was struggling into existence) he wrote to Balfour, partly to thank him for sending him a copy of Professor Graham's Catalogue,[24] and commented that 'one had come hither addressed to me, for [use by] the Botanical Socy. of London, which was accordingly forwarded to that 'scientific' body'. This

---

[22] Watson 1847–59, III, 4–5.
[23] For the situation in Britain as a whole, see Allen 1976, 103–14; for the Botanical Society of Edinburgh, see Fletcher & Brown 1970, chap. 10; for the Botanical Society of London, see Allen 1986, chaps 1–5.
[24] Probably Graham 1835.

was obviously a sneer. Wishing to give their Society some status, the London amateurs looked around and found for their president, not Watson, but J.E. Gray, a zoologist at the British Museum.[25]

Watson was 'much gratified' by the founding of the Edinburgh Society, and hoped it would flourish. On 26 June he thanked Balfour for plants sent, and asked what he wanted in return. He urged that it extend its scope to include exotic botany, and regretted that its species exchange program was not located in London. In 1841 Balfour invited Watson to accompany him and Professor Graham on an excursion into the Highlands and Hebrides. Watson expressed an interest in the trip (11 June 1841) and apparently went along, for in September he wrote:

> When in Edinburgh I asked to see your specimens of *Cnicus Fosteri*. They are exactly similar to the Weybridge specimens gathered by [John Stuart] Mill ... at the corner of Ditton Marsh. After my return from Scotland, I visited the locality which I supposed he meant, & there I found the dried remains of a branched *Cnicus*, apparently *C. Fosteri*, in the very spot where three years ago I gathered only *C. pratensis*. This wet summer, the latter has become (your & Hooker's) *C. Fosteri* there.

Besides provocative transmutational observations, Watson also included comments which seems related to the business of BSE:

> I presume Mr. [William] Brand has not overcome his 'indignation' against me, since he has taken no notice of letters which I sent to him, showing that it was his *red lines, not my censure* of the lithographer for map-inaccuracies that prevented the recommendation committee again granting money.[26]

Brand was a minor figure, but Watson was soon criticizing Balfour (29 January [1842]):

> In the Edinb. Cat. there is a sort of Hibernicism that I shall probably be wicked enough to illustrate in print. Example: '*Radiola millegrana*. Rare within 16 miles of Edinburgh.' Also, 'not found at all within 16 miles of Edinburgh.' (Professor Balfour, Edinb. Bot. Soc. Cat., 2d Edition)

---

[25] Allen 1986, 12–13. On Gray, see Gunther 1974 and 1975.
[26] On Brand (1807–69) and Mill (1806–73) as botanists, see Desmond 1977, 85 and 437.

By confessing his 'wicked' temptation, Watson showed some awareness that his negative responses to colleagues were not dictated entirely by a desire to maintain high scientific standards.

Balfour was amicable enough to accept such antics without expressing offense. In his next letter (25 February 1842) Watson thanked him for a testimonial in support of his application for the King's College Chair of Botany and then complained about the quality of the specimens currently disseminated by BSE. This might not have been a complaint against Balfour himself, as he was by then the Professor of Botany in Glasgow. Watson explained that BSL had begun sending out exchange parcels of specimens of better quality, and he was ready to give up on the Edinburgh exchange program.

While Balfour was in Glasgow, 1841–45, Watson's letters to him were mainly about exchanging specimens and their identifications. Balfour's return to Edinburgh, however, opened up new opportunities for complaints about how BSE was run (13 January 1846):

Thanks to you for the examples of *Poa Balfourii* – part of which shall be presented to the London Society's herbarium.

Sometimes I wish there was only one Botl. Socy. for exchanges. I became an *active* member of the London [one], on finding the Edinb. Socy. so uncertain & procrastinating that I could not get the British plants required – that is, the new & dubious things. From that time, I have felt increasing distaste to the Edinb. Socy. – partly from construing (it may be misconstruing) certain petty neglects into *intentional slights, because I had upheld the London Society.*

One function of this diatribe was to justify his shift in loyalty from the Edinburgh to the London society. There was really no need for an excuse, and the one he offered hurt his standing more than it helped.

On 18 August, when Watson had some hope of obtaining a chair of botany in Ireland, he wrote to Balfour and thanked him for his testimonial. He also asked a self-indulgent rhetorical question: 'How think you the Bot. Socy. London will get on, without me there to apply the spur or touch the reins?' Balfour may not have been willing to indulge Watson's ego by agreeing that BSL would fall apart without him, but Balfour did assure Watson that he was whipping BSE into shape. Watson responded with a little condescension and skepticism (30 December 1846):

I am glad to find the Edinb. Society shows some symptoms of revival in its distribution department. But I cannot say much in praise of my [received] parcel. It contains some few of the things wanted three years ago, but long since supplied from London. The nomenclature is very bad. The *Reports & Transactions* are well done, & the arrangements with the *Annals* is politic – at least for Mr Babington, as he thus obtains a veto on everything proposed for the Edinb. Socy's *Transactions*; that is, by refusing it for the *Annals*.

It is unclear how Balfour responded to these remarks, but this was Watson's last letter to him for eighteen years.

Chapter 10

History Not Quite Repeated: Watson, the Botanical Society of London and *The Phytologist*, 1840–58

The readers of *The Phytologist*, and all botanists, are much indebted to Mr. H.C. Watson for his careful, and I believe accurate investigation, in the January number, of the three species of œnanthe ...
John Stuart Mill 1845, 48.

Watson had tried some strategies on behalf of phrenology in the 1830s which were applicable to botany in the 1840s. He had advocated a grass-roots phrenology carried out by capable amateurs who were led by a few professionally trained scientists, such as himself and Andrew Combe. It had been unrealistic to expect Britain to support many full-time phrenologists. Professional leadership was to be exerted through textbooks, a semi-popular scientific journal, and a national organization that tied together local ones. The organizational ideas had failed because phrenology never became a creditable science. Botany, however, was already creditable, and periodicals were being published. Watson would not have to take up an editorship to keep botany going. About the same time that he advised George Combe on how to broaden the readership of *The Phrenological Journal*, he advised William Hooker about broadening the appeal of botanical journals. Although Hooker considered Watson's advice, he developed his own strategy for capturing different audiences with scientific and semi-popular journals. The challenge was to maintain subscriptions in the midst of a proliferation of competing journals. Hooker coped by editing, in whole or in part, an astonishing succession of journals for more than a quarter-century.[1]

---

[1] *Botanical Miscellany*, 3 vols, 1830–33; *Journal of Botany*, vol. 1, 1834, vols 2–4, 1840–42; *Companion of the Botanical Magazine*, 2 vols, 1835–36, n.s., 1845; *Jardine's Annals of Natural History*, 4 vols, 1838–40; *The London Journal of Botany*, 7 vols, 1842–48; *London Journal of Botany and Kew Garden Miscellany*, 9 vols, 1849–57; Lawrence et al. 1968, 420. Hooker also edited two periodicals which emphasized illustrations: *Exotic Flora*, 3 vols, 1823–27, and *Icones Plantarum*, 10 vols, 1837–54; Pritzel 1871–77, nos 4215 and 4224. For the general situation, see Allen 1976, 96–98, and Sheets-Pyenson 1981a.

By November 1840, Watson had given up his phrenological involvements, and although a member of BSE, he was nevertheless interested in seeing what he could make of BSL. Its founders, though not its nominal President, J.E. Gray, pursued careers which often took them away from the region. Watson was glad to become one of BSL's two Vice-Presidents, an honorary position which he could nevertheless exploit. BSL had been industrious in obtaining exchange specimens from its members, but it needed some one knowledgable and efficient enough to run its exchange program. Although he was dismayed at the chaos, Watson was the man for the job. Because the members were supportive and his leadership was uncontested, things went well for a time. BSL's historian explains that there was also another bit of good luck:

... the 20-year-old who had succeeded Chatterley as Secretary, and was thus the Officer to have to work with Watson most intimately, was perfectly suited to the role. George Edgar Dennes was meek, biddable, unassuming, conscientious to a fault: the natural task-master had inherited the ideal slave.[2]

There was, however, a disparity between the priorities of Watson and many, probably most, BSL members. His goal was to acquire a collection of specimens of each species of British plants in the different parts of its range in order to advance plant geography and to document morphological differences which might develop in response to environmental differences. He wanted to make a serious contribution to botanical science. The great majority of BSL members, however, seem to have joined because of a hobby concerning flowering plants and ferns. They usually enjoyed both field excursions and collecting dried specimens for a personal herbarium. It was an age of collecting: other naturalists collected sea shells or birds eggs, and a number of collectors also made drawings of plants, animals or shells.[3] These hobbies were pursued by women as well as men. The level of accuracy required by many such hobbyists was somewhat lower than Watson's. When his precision meant that specimens were expertly identified, BSL members were grateful, but when they became the target of his ire because they contributed specimens for exchange without following all of his instructions, their gratitude waned.[4]

---

[2] Allen 1985, chaps 2–3. Watson acknowledged Dennes's services by naming a new Azorean species in his honor: *Vicia dennesiana Wats.*; Desmond 1977, 181.

[3] A comprehensive and authoritative survey is Allen 1976; see also Allen 1969. A more popular treatment is Barber 1980. On British women who studied plants, see Allen 1980, May 1990 and Shteir 1987.

[4] The earliest instructions are undated but were probably printed in 1841; one copy is extant. A slightly modified version was printed in 1843; a few copies are extant in the British Museum (Natural History) and the British Library. Allen 1986, 31 and 183, n. 25.

Running BSL's exchange program was tedious and time-consuming, and Watson convinced BSL to hire a part-time curator at a modest annual stipend. Since the rewards were slight, and since Watson saw his own role as applying the spur and the whip, BSL used up a succession of three Curators in as many years. A fourth candidate wanted to supplement his pay by buying and selling specimens, but Watson thought that would be an unacceptable conflict of interest, therefore he and Dennes shouldered the drudgery for several years, with occasional voluntary assistance.[5]

Natural history periodicals appeared (and disappeared) almost yearly, and one appeared in London in 1841, *The Phytologist*, which met the needs of Watson and BSL. It began under the partnership of Edward Newman and George Luxford. Newman had published papers on insects in the 1830s, then in 1840 Luxford published Newman's *History of British Ferns*, which was a commercial success. This inspired them to publish *The Phytologist*, which was not a commercial success but nevertheless continued under their joint management until Luxford died in 1854.[6]

Balfour, Babington and W.H. Campbell compiled the second edition of BSE's *Catalogue of British Plants* in 1841. It served as a checklist for exchanging specimens among that society's members. In an anonymous review for *The Phytologist*, Watson recommended that its alphabetical arrangement 'be strictly adhered to; for we consider the systematic arrangement of species in a merely local list, to be a matter of minor importance'.[7] A few months later he changed his mind. He still conceded that 'it is absolutely necessary for distributing societies to have a fixed series of names, to facilitate the tedious processes of sorting their specimens, and of supplying the desiderata-lists of applicants; and the alphabetical order certainly offers many advantages for this purpose'. However, if one considered the needs of science rather than the practicalities of an exchange program, a systematic arrangement was far better. He spoke from experience: 'I am now engaged in preparing for the press a somewhat voluminous work on the distribution and localities of the plants of Britain, and have found it so very inconvenient to make references to alphabetical catalogues, or even to consult them, as to have felt strongly disposed to draw a line of distinction between the alphabetical and arranged lists; rejecting the former as unsuitable for the purposes of science.' That might sound capricious, but 'with one, or at most two exceptions, this line of distinction would throw out only the inaccurate lists of inexperienced and incompetent botanists'. Hints in his private correspondence indicated that the Edinburgh

---

[5] Allen 1986, 29–31; Watson to W. Hooker (10 December [1841?]).
[6] [T. Newman] 1876, 18–20.
[7] [Watson] 1841c. Watson also praised it in a letter to Babington (29 October 1841).

Catalogue was on his short list of competent alphabetical catalogues, but Watson did not say so in print, to avoid 'wounding the feelings of others without any necessity'.[8] Babington and Balfour might have noticed that Watson's new-found consideration for the feelings of others enabled him to avoid publicly endorsing their *Catalogue*. Watson easily persuaded BSL to sponsor a rival, systematically-arranged list compiled by him, assisted by Dennes. *The London Catalogue of British Plants* first appeared anonymously in 1844; its seventh edition appeared in 1874 (with further changes likely in the 1877 and 1881 printings). *The Edinburgh Catalogue*, in contrast, went through only four editions, the last being in 1865.[9] Although the *London Catalogue* probably won this competition on its merits, it was given an initial boost in a lengthy anonymous review in *The London Journal of Botany*, which was edited by William Hooker. The echo of Watson's sentiments and the praise of his works indicate that he probably wrote the review.[10]

It did not occur to Newman that Hooker would let Watson review anonymously his own *London Catalogue* in the pages of *The London Journal of Botany*, therefore Newman blamed Hooker for the biting attack contained in the review on some nomenclature used in the second edition of Newman's *History of British Ferns*.[11] Newman was also misled by the fact that Watson had reviewed it favorably in *The Phytologist*.[12]

In 'Remarks on the Distinction of Species in Nature, and in Books' (1843) Watson took to task the authors of various Floras of England – Hudson, Withering, Smith, Lindley, Hooker and Babington – for their inability to agree on delimitations of British species. It must have been the appearance of Babington's *Manual* which inspired his comments. Watson approved of adding the name of the first describer of a species after the species name to indicate the authority for the name, but he objected to a later substitution of another botanist's name who renamed the species:

> Those who can discover nothing themselves, or who can get no novelty to describe, are, at any rate, able to divide and re-name the genera or species already well-known; and by so doing, they can acquire a sort of spurious claim to put the abbreviation of their own names along with those of the re-named plants. Hence

---

[8] Watson 1842d. The 'somewhat voluminous work' to which Watson referred was a third edition of *The Geographical Distribution of British Plants*, the first and only part of which he published in 1843 (it was on too large a scale to be continued).

[9] Freeman 1980, nos 189 and 3917.

[10] [Anon.] 1844.

[11] [E. Newman] 1844. Yet he or Luxford had earlier allowed Watson to announce anonymously the pending publication of The *London Catalogue* in their own periodical; [Watson] 1844e.

[12] [Watson] 1844f.

come the rapid increase of false species, and the confusing subdivisions of well-defined and long-recognized genera.[13]

Watson had a genius for finding real problems which had inadvertently arisen, and then imputing unworthy motives to whoever got caught up in the dilemma. Botanists operating on the assumption that species are stable usually imagined that a particular change in names which they thought needed was a minor problem amenable to a commonsense solution. Watson, who believed in evolution, saw that the dilemma was not minor and that commonsense solutions were inadequate. However, instead of proposing that formal rules of nomenclature be drawn up by an authoritative committee, he was content to cast blame. He concluded dryly that, 'fortunately, [these name changers] cannot change the species in nature also, if permanently distinct species do certainly exist'.[14] This remark may contain a clue as to why he did nothing but complain: if species are changeable, there might not be any permanent solution to name problems.

That unsettling possibility might have caused some to throw up their hands in despair, but not Watson. He wanted to see how precisely species could be defined, and his method of doing so was to harass those involved in the social enterprise of descriptive botany until they achieved his own level of precision. This approach resembles his unsuccessful harassment of phrenologists in hopes of making them into more sophisticated investigators. At least descriptive botany was on a more realistic footing than descriptive phrenology. Even if he could not coerce those who believed in unchanging species into seeing nature as he saw it, he could often demonstrate his findings to them well enough to gain acceptance. One example is his separation of specimens commonly lumped under the names of two species of water dropworts, *Œnanthe pimpinelloides* L. and *Œ. lachenalii* Gmel. He demonstrated his specimens at a BSL meeting, explained the distinctive characteristics of each species, and showed specimens of a third species which presumably Sir James Edward Smith identified as *peucidanifolia* and John Ball as *silaifolia* Bieb.[15] Lacking access to the original specimens of the latter two species and mistrusting the latter two judgments, Watson gave his third species a temporary name, *Œ. Smithii*.[16] However inept Ball may have been in other pronouncements, on this one he seems to have been correct, as Watson later realized.[17] *Œ. silaifolia* Bieb. is rare in southern England and the Midlands and specimens collected had probably been conflated with one or the other (or both) of the above species, not only on herbarium sheets, but in the published record, as Watson found.

---

[13] Watson 1843c, 615.
[14] Watson 1843c, 622.
[15] Ball 1844. On John Ball (1818–89), see Desmond 1977, 33.
[16] Watson 1845b.
[17] Watson 1845e. Cf. Babington 1844a and 1844b.

John Stuart Mill was so grateful to Watson for clearing up this confusion (to the extent that he had) that he published a glowing testimonial. Watson obligingly examined Mill's specimens and verified their identies. Mill then sent in another note to *The Phytologist* thanking him.[18] Watson appears here in a positive light because his judgments were welcomed; Mill was not one of those whose identifications Watson called into question. It is always easier to applaud the advancements of science when they have not been made at the expense of one's reputation.

Meanwhile, not all botanists who read *The Phytologist* and used *The London Catalogue of British Plants* had witnessed Watson's authoritative presentations of specimens at BSL meetings. Provincials sometimes imagined that it was safe for them to criticize his work in a democratic spirit of give and take. One of these was Joseph Sidebotham, a calico printer and a founder of the Manchester Field Naturalists' Society, who welcomed use of the natural system for listing names in BSL's *Catalogue* but thought the names should be the same as those in BSE's *Catalogue*. He itemized his many objections, such as: 'Surely *Prunus insititia* and *domestica* are something more than varieties of *P. spinosa*!'[19] Newman incautiously added his own nomenclatorial thoughts to Sidebotham's as an editorial note. Watson wrote four responses to these criticisms, under both his own and Dennes's names. Although his critics questioned some of Watson's judgments, they did not attack his competency. His substantive responses were interspersed with progressively more disrespectful comments when critics persisted. To Sidebotham's initial criticisms, his responses were merely impatient: '*Impatiens fulva* was originally introduced to this country from America, as any botanist would have informed Mr. Sidebotham.'[20] However, since he did not deign to respond to Sidebotham's full list of queries, the door was open to further complaints, this time from Sidebotham's Manchester colleague, Leo Grindon.[21] Watson began with a counter-complaint:

> Some correspondents of *The Phytologist* (pp. 972, 1077) have been inditing censures, in the disguise of queries, on the nomenclature of the *London Catalogue of British Plants*, compared with that of Edinburgh. An importance is given to those censures, which they would not in themselves possess, by an editorial note (p. 974) being attached to one of the letters.[22]

---

[18] Mill 1845; part of this note is quoted at the head of this chapter. On John Stuart Mill (1806–73) as a botanist, see Desmond 1977, 437; on Mill's personality, see Mazlish 1975.
[19] Sidebotham 1844. On Sidebotham (1824–85) as botanist, see Desmond 1977, 559.
[20] [Watson] 1844h, 1014–15.
[21] Grindon 1844. On Leopold Hartley Grindon (1818–1904), see Desmond 1977, 272.
[22] Watson 1844l, 1128.

Because Watson distorted both the letters and the editorial comments in his replies, Newman added a further note to Watson's response stating that Watson appeared to misunderstand parts of both the letters in question and his own earlier editorial comments. When BSL issued a second edition of *The London Catalogue*, Sidebotham resumed his criticisms, and Watson replied tit for tat, until Newman mercifully ended this exchange within the pages of *The Phytologist*.[23]

There were enough members of BSL who shared Mill's gratitude toward Watson's botanical investigations and services to the Society that in 1846, 68 of them contributed funds to have portraits painted of him and Gray, and later one painted of Dennes. Their portraits by Margaret Carpenter were presented to them at the Anniversary Meeting, and they returned them to the society to hang in its Bedford Street meeting room.[24]

BSL members undoubtedly assumed that their esteem for Watson was mutual, but it was not (and he nevertheless sat for BSL's portrait without feeling it inappropriate to do so). He confessed his low opinion of BSL and of Dennes in letters to Balfour on 13 January and 18 August. He suspected that BSL would collapse if he went to Ireland, and he may have persisted in office as long as he did because he imagined that he would soon follow the earlier officers in moving away to a new career. That hope slipped away, and early in 1849 he announced his desire to resign as Vice-President, but he continued on as a dominant member until BSL disbanded.[25] He wanted to reduce BSL to an exchange club, without the encumbrance of library, herbarium and meetings. The Linnean Society could perform these functions for London. His idea ignored the differences between the two organizations – the Linnean Society was open to those who were elected on the basis of scientific achievement or because they were wealthy amateurs, while BSL was open to anyone who paid its modest dues.[26]

Watson may have enjoyed some of his controversies, but he also had doubts. In a letter to Combe on 25 August 1846, he justified his outbursts:

The excess of [inherited?] Destructiveness is a serious blemish. The gunpowder is there, [igniting] at the chance of a spark. But I write much less ferociously now, than ten years ago. And [I] very rarely get angry in oral discussion. This is not [a] real improvement, but the contrary.

---

[23] Sidebotham 1848a, 1848b and 1848c. Watson 1848m. Newman's ending of the debate was justified on 189, following Sidebotham 1848c.

[24] On Watson and Gray's, see Allen 1986, 37 and 184, n. 38. Allen's book also reproduces both portraits, frontispiece and 23. On Dennes, see Reynolds 1847.

[25] Allen 1986, chaps 3–7.

[26] Allen 1986, 37–38. Gage & Stearn 1988.

Yet, some six weeks later (4 October) he admitted that loss of temper could be counter-productive:

> Some furious onslaughts on anti-phrenologists, and the necessity of strangling an impertinent botanist or two, and showing some others that they had better let me alone, have got me the credit of being very pugnacious; a credit which I do not wish to increase needlessly.

He hoped to minimize damage to his reputation by publishing some of his attacks anonymously. In 1849 James Backhouse Jr called into question this tactic after being attacked by 'C'. *The Phytologist* published his retort immediately following one from Babington, who was responding to two complaints signed respectively by 'C'. and H.C. Watson.[27] Backhouse seems to have already made the connection between these two signatures before his and Babington's responses appeared together:

> I would recommend that if 'C.' (or any writer) gives another paper on this or any other subject, he bring out his name openly, as my previous 'strictures', 'false witness', &c. would probably have been spared [appearing in print] had he done so before. In a journal like *The Phytologist* no one ought to suppress his name, to make remarks in secret which he would *shrink* from making openly.[28]

Neither chastisement nor exposé had an immediate effect – Watson published in *The Phytologist* under his 'C.' signature five more times in 1849 and 1850 – but his ceasing to do so after that was possibly related to Backhouse's strictures. The editors of *The Phytologist* might have declined to take any more of his anonymous contributions, for he stopped publishing in their journal. His later sour comments about the new series of *The Phytologist* (May 1855–July 1863) may have had as much to do with his personal conflict with its editor, Alexander Irvine, as with its alleged decline in scientific standards.[29]

---

[27] Babington 1849b. See also Babington 1849a.

[28] Backhouse 1849. By 'my previous "strictures", &c.', he meant those which had been directed toward him by 'C.' He was responding to [Watson] 1849g. On Backhouse (1794–1869), see Desmond 1977, 26.

[29] '[I] ignore the new series of that Periodical, as now (1855–8) scarcely belonging to the category of scientific publications. A periodical printed for the purpose of sale, as a trade speculation, cannot be conducted on the same strict principles as one printed for the promotion of science. To unite the two objects, by intercepting blunders and twaddle, without repelling subscribers, would require a very competent and judicious Editor.' Watson 1847–59, IV, 4. On Irvine, see Boulger 1891 and Desmond 1977, 337. On the friction between Watson and Irvine, see Allen 1986, 36.

Watson's desire to curtail the scope of BSL's activities, his unwillingness to continue serving as its Vice-President, and the sharp decrease in his publications in journals were probably related to his having embarked on his *magnum opus*, entitled *Cybele Britannica; or, British Plants and Their Geographical Relations* (4 vols, 1847–59 + later supplements). By 1850 he realized that he would have to cease other publishing to ensure that he could finish this one in a reasonable time.

During the summer of 1851 BSL responded to Watson's reduced involvement by hiring another curator – after six or seven years without one. Watson recommended John Thomas Irvine Syme (later Boswell-Syme; still later, Boswell) (1822–88), the son of an Edinburgh natural history artist, who came with strong BSE recommendations.[30] Initially, Syme was a strong asset, but BSL's problems extended beyond Watson's disengagement. Dennes suffered financial reverses which seriously interfered with his secretarial duties. His troubles were due in part to his neglect of his law profession for his botanical hobby; the result was disaster in both areas. Syme supplemented his income with various other botanical jobs, and he may have been too busy to notice Dennes's deficiencies until it was too late. At the annual meeting on 24 November 1856 Watson, Syme and John Reynolds, the Treasurer, convinced the membership that Dennes had so mismanaged finances as to leave them no practical choice but to disband. BSL's herbarium was auctioned off on 9 February 1857, overdue rent and other expenses were paid, and the remaining funds were presented to Dennes, who by then was 'in a starving condition'.[31] (The three portraits were probably returned to their subjects; Watson's was eventually hung at the Royal Botanic Garden, Kew, J.E. Gray's at the Royal Society of London, and Dennes may have taken his with him to Australia.)

Watson was not deeply disappointed at this turn of events, as it opened the way for substituting an exchange club for BSL, as he had long since suggested; it would mainly attract scientifically minded botanists. *The Phytologist* still existed to meet the needs of hobbyists, who organized the London Natural History Society in 1858. Serious London botanists who wanted more than an exchange club could attend meetings of the Linnean Society.[32]

---

[30] Desmond 1977, 78.
[31] Allen 1986, chap. 5; quotation, 61: a letter from Watson to John G. Baker, 19 January 1866.
[32] Allen 1976, 163; 1986, chaps 5–6. Gage & Stearn 1988, chap. 4.

Chapter 11

The Origin and Transmutation of Species, 1832–47

> So little are species distinguishable, so liable are some of them to run into variations, that no botanist can now tell what are distinct species of rose, bramble, willow, mouse-ear, sedge ...
> Watson 1836b, 27

Watson was one the first scientists to develop and carry out an original research program to obtain data showing that evolution is occurring, though it was an implicit rather than a published program. Erasmus Darwin and Lamarck certainly collected evidence for their discussions, and Lyell collected evidence in order to discredit Lamarck's ideas. However, it seems doubtful that their efforts could be called developing and carrying out a research program in the same sense that Watson did. For the most part, these scientists were compiling and organizing existing information in order to build their arguments. Watson undertook original research to document evolution. This was also true of zoologist Étienne Geoffroy Saint-Hilaire by the 1810s, but his research program might also have been implicit, like Watson's, until much later.[1]

Watson's Orientation and Studies

Edinburgh was where transmutation reappeared after Erasmus Darwin's uninfluential beginnings in the 1790s. Robert Jameson, the prominent Professor of Natural History at the university, published an anonymous defense of the idea in 1826, and his student, Robert E. Grant, espoused Lamarckism to Charles Darwin before departing in 1827 for the University of London.[2] One might therefore suppose that Watson adopted the idea while living in Edinburgh. That supposition seems wrong, however. In his *Outlines of the Geographical Distribution of British Plants* (1832) – published after he left Edinburgh – he acknowledged that

---

[1] I follow Bourdier 1972 on Geoffroy Saint-Hilaire's career and writings.
[2] Desmond 1984; Secord 1991.

'investigations concerning the original creation of plants, in the [present] state of human knowledge, might be deemed by many at best an idle waste of time'. Some naturalists thought that all species come from the same place, others thought there were several centers of creation from which species spread, and still others thought that they originated where they now exist. 'Some again suppose, that at first only *genera* existed, *species* arising from generic admixture; whilst others maintain that all vegetable forms are modifications of each other, or the result of a certain concurrence of molecules dispersed through matter; hence liable to be produced in any situation where the necessary conditions of their existence occur.'[3] Intrigued though he was with notions on transmutation, he was noncommital.

His earliest known commitment to the notion is in a letter to Nathaniel Winch (7 October 1834) explaining his ideas on the distinctiveness of species within difficult genera:

> As to [John] Lindley's brambles: – I look on it that any genus might be divided into many species, if such small differences are to be held specific, and that such divisions should be made in all or none, to keep uniformity. *Species* in any sense or degree I look on as human divisions, not as the creations of nature. The changes, proved by geologic evidence, to have occurred in organic forms, and those now effecting by climate, elevation, cross-breeding, &c. &c. strongly discountenance the idea of absolute and permanent distinctions.

The problem at hand, how to separate a batch of brambles, was concrete but recalcitrant. Watson's skepticism about there being any definitive solution to it smacks of Lamarckianism: Lamarck did not accept the reality of species.

Although Jameson and Grant were influenced directly by Lamarck's writings, most British naturalists learned about Lamarck's theory from Lyell's lengthy attack on it in *Principles of Geology* (vol. 2, 1832). A number of younger naturalists were not convinced that Lyell had had the last word on the subject,[4] and Watson was one

---

[3] Watson 1832a, 2–3.
[4] Herbert Spencer wrote (1904, I, 201): 'my reading of Lyell, one of whose chapters was devoted to a refutation of Lamarck's views concerning the origin of species, had the effect of giving me a decided leaning to them. Why Lyell's arguments produced the opposite effect of that intended, I cannot say. Probably it was that the discussion presented, more clearly than had been done previously, the conception of the natural genesis of organic forms.' Darwin, who had previously read some of the writings of both his grandfather and Lamarck, read Lyell's *Principles* during the voyage of the *Beagle* but still did not become an evolutionist until shortly after his return to England. Wallace read Lyell at about the time of his conversion to evolutionism but attributed the conversion to his having also read [Robert Chambers's] *Vestiges of the Natural History of Creation* (1844); Schwartz 1990, 140–7.

of these. In the year or two since publishing *Outlines*, he either read Lyell or talked with naturalists who had. Transmutation was one of those radical notions – phrenology was another – which suited Watson's temperament, and it would have been surprising if he had not accepted it.[5] Oddly, however, Watson the botanist was always more cautious than Watson the phrenologist. His earliest published support of evolutionism was thus in a phrenological rather than a botanical context.

The occasion he chose did not require the use of transmutation theory, but since it was now part of his intellectual arsenal, he chose to make it relevant. William Scott's lengthy polemic against Combe's *Constitution of Man* prompted Watson to publish a rebuttal. While preparing it, he warned Combe (4 September 1836) that his refutation of Scott would 'disagree with your mode of explaining the laws of nature, altho' hardly differing in respect to their reality & practical consequences ... You read nature, see certain events, and infer those events to be arranged by a Creator & Designer. Science, however, teaches only things that exist & occur, and their relations. It does not shew a Creator or a Superintendent.' Combe asked Watson to proceed with it, despite their philosophical differences. Watson's rebuttal would have attracted few readers outside the phrenological movement, and even those within it would have had to read some twenty pages of Scott's misrepresentations of Combe before encountering Watson's evolutionary views. Scott had asserted that

> When a new species of plants or animals appears to have been created, there is nothing like gradation or progression. The new species is not derived from an older and more imperfect one, but starts at once into existence, at the Almighty fiat, in all its completeness and perfection.[6]

Watson's rebuttal shows familiarity with Lyell's *Principles of Geology*, and it was among the works which he recommended as an antidote to Scott's *Harmony*. Although some of Watson's arguments were Lamarckian, he never mentioned Lamarck. Lyell's discrediting of Lamarck made mention of his name inexpedient. Scott's anti-Lamarckian argument (originally from Cuvier), that mummified animals from Egypt show that species do not change was mooted by Watson's retort that 3000 years did not constitute an adequate test, especially in a land with a stable climate. Furthermore, the existence of new and better breeds of dogs and other domestic animals which were unknown in antiquity was evidence for transmutation. Watson thought that: 'if man is ever to create a permanent species, he must go to work in a much more gradual manner, by coupling together varieties becoming more and more unlike the original stock at each descent'. For those wishing to undertake

---

[5] Cf. Desmond 1985, 1987 and 1989.
[6] Scott 1836, quoted from Watson 1836b, 22.

such investigations, 'our gardens now abound with intermediate varieties or transition-species'.[7]

Combe wanted to know more than Watson had revealed in his counter-attack on Scott about the history of the earth and its species, and Watson sent him (16 December) a brief account of geological change and the fossil record. The latter seemed to indicate a progression in animal life, but this might be only an illusion:

> as the inferior types have far more species and individuals than the superior ones, and as the highest types inhabit the land chiefly (birds amd quadrupeds), the probability is that thousands of the inferior may have been fossilized for each one of the superior animals; and deposits are doubtless now forming in the sea, which may not have the bones of one bird or quadruped for a million of shells or fish-skeletons, and thus would become geologic evidence, negatively, that birds and quadrupeds did not coexist with them, unless the one happened to be formed. The evidence apparently in favour of progressive improvement in the kind of animals is thus only indirect, and may possibly be annihilated at once by the discovery of a mammalia in the oldest rocks containing vestiges of life.

Watson was uncertain about whether nature could improve itself continuously – 'the evidence only shews oscillation to and fro' – but the historical record indicated that humanity improves both itself and nature (the latter by breeding domestic species). If he persisted in his uncertainty about whether irreversible changes occur in the wild until after he read Robert Chambers's anonymous *Vestiges of the Natural History of Creation* (1844), there is an interesting possibility that he and Chambers influenced each other's conversion to evolutionary belief.

When Watson stated in 1836 that 'our gardens now abound with intermediate varieties or transition-species', he had definite examples in mind: 'rose, bramble, willow, mouse-ear, sedge, and many others ...'. He did not follow up on his own hint, however, because he soon dropped botanical research to edit *The Phrenological Journal*, 1837–40. When he resumed botanical studies, his first publication was a 'Description of a Primula, found at Thames Ditton, Surrey, Exhibiting Characteristics Both of the Primrose and the Cowslip'. He was recalling attention to something earlier botanists had noticed: England has two common primulas: Primrose (*Primula vulgaris* Huds.) and Cowslip (*P. veris* L.), and also Oxlip (*P. elatior* (L.) Hill) of more restricted range; Primrose has relatively large (2–3 cm wide) pale yellow flowers which arise on single stalks from the rootstock; Cowslip has relatively small (1–1.5 cm wide) pale yellow flowers which grow in a terminal cluster. In 1819 William Herbert had collected seeds from a Cowslip, which may

---

[7] Watson 1836b, 26–7.

actually have been a hybrid, for the plants raised from them were highly variable and resembled Primrose, Cowslip and Oxlip. John Stevens Henslow was skeptical of this until he found at Westhoe, near Cambridge, 'in great plenty a peculiar variety of Primula, which I scarcely knew whether to call the oxlip or the cowslip'. He then examined more closely the Cowslip in his garden and found it had changed character over the period of a year, until it came to resemble Oxlip. Both botanists's observations were of only a few specimens and under informal conditions. Nevertheless, in *Principles of Geology* Lyell cited their conclusions as evidence that the existence of varieties in nature is no evidence for the transmutation of species.[8]

Whether or not Watson had read and remembered these earlier accounts, his control of growth conditions were no better than Herbert's and Henslow's. While growing Primrose and Cowslip in his garden, he discovered an Oxlip among them. His note simply reported: 'All the circumstances lead to a reasonable presumption that this solitary oxlip had originated from a seed either of the cowslip or primrose; yet its characters are so completely intermediate between the two, that I can give only the slightest preponderance in favour of the cowslip.'[9] He did not guess how it had arisen. His note prompted Henry Doubleday from Bardfield to send him specimens of local Oxlip, which Watson judged:

to be the species intended by the figure in [James Sowerby's] *English Botany* (513), and also to be identical with Swiss and German specimens in my herbarium, which were sent to me under the name of '*Primula elatior* Jacq.' The Bardfield specimens differ slightly from the figure in *English Botany*, but not importantly, except in having the calyx decidedly shorter than the tube of the corolla. They are unlike any other English oxlips in my herbarium (all of which may be gradually traced either to the primrose or to the cowslip, by intermediate links), and, as appears to me, they may be safely pronounced the real representatives of *Primula elatior*.[10]

Watson was unable to find any of the supposed specific characters given in the manuals for primrose and cowslip which were constant, and he encouraged botanists to clarify this matter by careful breeding experiments. Two years later he found a

---

[8] Herbert 1822. Henslow 1830. Lyell 1830–33, II, 34–5. On all three as botanists, see Desmond 1977, 303 and 399. On Herbert, see also Guimond 1972; on Henslow, see also Russell-Gebbett 1977; on Lyell, see also Egerton 1966, 232–40, and Wilson 1973.

[9] Watson 1841a. For modern studies on the taxonomy and genetics of these primulas see Valentine 1947, 1948, 1952, 1953, 1955, 1961 and 1966.

[10] Watson 1842g. Pritzel 1871–77 (no. 8789) lists *English Botany* under Sowerby, the illustrator, but Freeman 1980 (no. 3455) lists it under James E. Smith, the sometime author of text.

trait in *P. vulgaris* and *P. veris* which was lacking in *P. elatior*, by which means he hoped to separate the last-named species from the spurious oxlip.[11]

In 1841 he had planted roots of the Claygate Oxlip in his garden, and in 1843 he planted some of its seeds in a pot. In 1845 there were 88 plants which he identified as follows:

True cowslips (*Primula veris*)	4
Cowslips passing to oxlips (*P. veris*, var. *major*)	5
Oxlips (*P. vulgaris*, var. *intermedia*)	23
Caulescent primroses (*P. vulgaris*, var. *caulescens*)	18
True primroses (*P. vulgaris*)	20
Plants not bearing flowers	18.

Since he had neither Cowslip nor Primrose growing in his kitchen garden, where he had his Oxlips, he felt that the plants of the latter would not have received foreign pollen 'unless through the agency of bees'. Although he continued this study until 1848, his methodology lacked rigor. He found what he had suspected all along – that Primrose and Cowslip produce a hybrid similar to, but distinguishable from, Oxlip.[12]

To collect plants, Watson went on field trips year after year, throughout England, Scotland and Wales. He also obtained and studied great numbers of herbarium specimens collected by others. He began doing so during the 1830s. In the 1840s he assumed control over much of the trading, lending, and purchasing of specimens, always for his own objectives, while also serving others, as he explained to William Hooker (10 December 1841? no. 416):

> I have promised to examine & label a large stock of duplicates for distribution in the hands of the Bot. Soc. of London – amounting probably to 20,000, & the mere labour of writing so many tickets is no joke.
>
> This may seem much trouble to undertake gratuitously; but it meets my two favorite departments of botany, namely, the study of localities & geographic ranges, and also the fancy of ascertaining to what extent species may vary in their habit & characters.

---

[11] Watson 1844g. Unfortunately, the printer misread the key term in his diagnosis and it was two years before he corrected it (P 2 [1846] 527–8). The diagnosis, with the error in italics and the correction in brackets was: 'In the cowslip and primrose, and all their varieties, a circle of scale-like *glands* [folds] surrounds the orifice of the tube of the corolla. These *glands* [folds] are absent from the *Primula elatior*.'

[12] Watson 1845h, i and 1848n.

Watson's arguments for the transmutation of species involved, first, pointing out certain genera about which there was no concensus on how many species and varieties grew in Britain, and then attempting to show that the basis for uncertainty lay in nature, not in the biases of botanists. In one discussion he mentioned his general conclusion that there were some 200 spurious species being accepted by British botanists.[13] When writing to Babington on 29 October 1841, he explained the slant of his current research, but without explaining that he wanted to bolster arguments for transmutation:

> I have latterly been looking out for any variations from the typical characters of species, in the belief that we should much better understand plants by bringing together series of specimens of each species, to illustrate the extent to which each one may vary. For example, *Linaria repens* and *vulgaris* in their typical forms are very little alike, save in general habit; yet I have specimens running so much from one to the other, as to render it dubious to which species some of the forms should be assigned. You will find examples of these in the parcel.

Two years later he announced to Babington (2 November 1843) his impression that 'no eternally permanent species exist,' while also admitting that 'this being the unproved, & unproveable impression of an individual only, I do not found anything upon it'.

Watson's opportunity to defend transmutation openly to botanists came in 1844 after publication of *Vestiges of the Natural History of Creation*. This sensational bestseller was written by Robert Chambers (1802–71) and published anonymously. Previous writings on evolution by Erasmus Darwin, Lamarck, and various minor authors had not attracted a large British audience. Vestiges provided the first detailed exposition for many readers. After reviewing it, Watson expounded his own ideas. His own commitment to continuous evolution in nature, as opposed to mere fluctuations, might have been encouraged by reading *Vestiges*. On the other hand, there is a good chance that Chambers's undertaking of his project was encouraged by Watson's *Examination of Mr. Scott's Attack upon Mr. Combe's 'Constitution of Man'* (1836). Chambers was a minor disciple of the Combe brothers,[14] and their respect for Watson's scientific judgment may have encouraged Chambers to read Watson's tract.

Be that as it may, Chambers was a publisher and writer whose scientific interests did not include botany, and he had not been reading Watson's discussions of species

---

[13] Watson 1844d; also 1841(?)e, 1842c, 1843c, 1843m, 1844b, 1844e, l.

[14] Millhauser 1959, 29. Some of the Chambers-Combe correspondence is quoted in Gibbon 1878; see 'Chambers' in the index.

variability published in *The Phytologist* and the *London Journal of Botany*. In a lengthy serialized discussion for *The Phytologist*, Watson offered some praise: 'Vestiges has exactly the character and qualities which are required in a really "popular work". The style is remarkably good and readable – the subject is great and interesting ...'.[15] However, as he wrote to Balfour (9 March 1845), 'the botanical part is quite a failure'. He was curious about who had written it. Harriet Martineau had it on good authority that it was written by Hewett C. Watson. She 'never believed it was [by] Robert Chambers, as the Carpenters declared'.[16] She obviously did not read *The Phytologist*, for Watson's critiques of *Vestiges* were signed.

Among the implausible evidences which Watson quoted from *Vestiges* was W. Weissenborn's report that under certain conditions the oats which are sown turn into rye before the harvest. Equally incredible was the claim that when a forest burns and different species then grow up, these plants are evidence of 'a progression of species which takes place under certain favouring conditions'. Watson responded that if these were verifiable facts, then:

> they overprove his theory. The change of the oat into rye is a pretty wide generic leap. And I am not at all aware that a burnt forest is forthwith succeeded by trees nearest allied, in specific or generic characters, to those which have been destroyed. The phenomena are here scarcely those of 'a simple and modest character', or an advance 'from one species only to another'. Had we been told that *Avena strigosa* could be so converted into the *Avena sativa*, or that a burnt forest of *Tilia parvifolia* would be succeeded by another of *Tilia europaea*, the changes would have corresponded better with the theory.

He ended his review with the comment that the author of *Vestiges* was 'slenderly acquainted with Zoology, and still less conversant with Botany.'[17] Despite the implausability of Weissenborn's claims, both Watson and Joseph Sidebotham repeated his experiment and reported their negative results to *The Phytologist*.[18]

Having rejected evidences in *Vestiges* concerning 'the origin and transmutation of species' (Watson's phrase), he provided his own in three continuation pieces: the first expounded general evidences, the second, specific evidences, and the third drew conclusions. He began with two questions which *Vestiges* raised: 'Can plants originate from unorganized matter?' and 'Can plants of one species, in any way,

---

[15] Watson 1845f, 109. See also Chambers 1994 and Hodge 1972.
[16] Harriet Martineau to James Martineau, 13 June 1845. I thank Roger Cooter for the quotation.
[17] Watson 1845f, 112–13. The quotation on Weissenborn's report is from [Chambers] 1844, 221.
[18] Sidebotham 1846. Watson 1846q. Cf. Weissenborn 1838a and 1838b.

produce individuals of another species?' Biology could not as yet answer the first of these. Nevertheless, he stated, it would be rash to assert that 'the simplest forms of vegetable life (say, for example, a Protococcus) never come into existence, unless by the development of germs which have first constituted portions of a parent individual similar to themselves.'

To answer the second question, there were three groups of evidence to consider: paleontological, the tendency of species to vary, and the existence of transitional forms between species. Since it had become clear from studies of the fossil record that species in past ages were different from those now living, he argued that there was no scientific explanation that could explain the facts except transmutation. Yet 'the supernatural alternative is the one generally received by the vulgar, and admitted – tacitly, at least – by men of science'. Although 'the stone tablets of Geology' made transmutation seem inevitable, he doubted that these tablets would ever reveal how it happened.[19]

Concerning the tendency of species to vary, he remarked that 'philosophical thinkers' had already concluded that the higher categories under which species are grouped – orders and genera – are mere conveniences. He was himself skeptical about the reality of species. His evidence in support of this skepticism was the inability of botanists to agree upon the number of species within Britain for a dozen genera: *Salix, Rosa, Rubus, Mentha, Viola, Festuca, Poa, Saxifraga, Cerastium, Hieracium, Polygonum, Myosotis*, and still others unnamed. The tabulation from six authorities in Table 11.1 illustrates his claim.

Table 11.1 Watson's comparison of claimed species from six authorities

	Salix	*Mentha*	*Rosa*	*Rubus*	*Saxifraga*
Hudson (1791)	18	6	5	5	9
Smith (1824–28)	64	13	22	14	25
Lindley (1835)	29	9	17	21	24
Hooker (1842)	70	13	19	14	16
Babington (1843)	57	8	19	24	20
London Catalogue (1844)	38	8	7	34	16.

The *London Catalogue*, of course, represented his own judgments. All of these examples were of wild species. When horticulturists chose to 'increase and extend

---

[19] Watson 1845f, 141–2. The first of these four parts, being his review of Vestiges, appeared in March, while the three parts with his own evidences appeared in April, May and July.

their variations, scarce any limit can be set upon the power of doing so', as seen in *Pelargonium, Erica, Rosa, Fuchsia* and *Calceolaria*.

That great variations can occur within a species must be conceded by everyone. The crucial question was whether such varieties ever produced new species. He had evidence which, although not proving transmutation, did at least cast doubt on the distinctiveness of species. *Geum urbanum* and *G. rivale* were accepted as distinct, yet intermediate forms were common, and there was no agreement over whether *G. intermedium* is a variety or a species. 'Apparently, both species sport into varieties; and these varieties run so near together as to have been combined into one supposed third species.'[20] Furthermore, the hybrids between *Primula vulgaris* and *P. veris* rather resembled the intermediate species *P. elatior* and are not sterile, as 'true mules' are.

His next installment in the series discussed the status of pairs of species within six genera and reputed changes among three orchidaceous genera. These six pairs were: *Viola canina* L. and *V. flavicornis* Smith, *Polygonum maritimum* L. and *P. Raii* Bab., *Lolium perenne* L. and *L. multiflorum* Lam., *Primula veris* L. and *P. vulgaris* Huds., *Festuca pratensis* Huds. and *F. loliacea* Huds., *Tolpis umbellata* Bert. and *T. crinita* Lowe. He had observed all of them in the field and in his garden. His garden reports will not inspire confidence in a modern reader because he only planted one or a few specimens of each and did not take precautions to control pollination, even though the pollinating action of wind and insects was common knowledge. The published reports which he cited of changes among flowers of the orchids *Monachanthus viridis, Myanthus barbatus* and *Catasetum tridentatum* surely would have aroused his own skepticism if he had been on the other side of the argument. In all these cases of orchids and species pairs in other genera he felt that the species were unstable and that specimens of one species could change over to the other when environmental conditions changed.[21] He did not present these conclusions as positively established. Experimental biology was not well developed in Britain, and was only slightly better developed in France and Germany. The casual methodology which Watson and various others used in experimentation would not have been accepted by him in herbarium research. Herbarium research was well established in Britain since Sir James Edward Smith had purchased the Linnaean herbarium in 1784 and helped found the Linnean Society of London in 1788.

There are, according to Ernst Mayr, three species concepts which students of nature have espoused: Plato's *essentialist* concept, Occam's *nominalist* concept, and Darwin's *biological* concept. Logically, these concepts are mutually exclusive, but in practise Mayr found that Lamarck fell between the nominalist and biological

---

[20] Watson 1845f, 145. Cf. Gajewski 1959.
[21] Watson 1845f, esp. 162–3.

concepts.[22] Watson's species concept may never have been precise, and may have changed from time to time. In his letter to Winch (7 October 1834), he seemed close to Lamarck; just over a decade later his concept seemed to be between essentialist and biological:

> ... plants repeat their own images by hereditary descent through a long series of years, to which we can assign no limit.
>
> These images, it is true, are not always perfect likenesses. Variations of climate and soil, or of other conditions, are accompanied by corresponding variations in the plants. But ... these variations are usually found to be temporary; so that we may say, there is a standard or average type for each kind, which is repeated in the individual plants as nearly as internal health and external conditions will allow. This supposed standard or average I will here express by the term 'central type'. ...
>
> Individual plants which differ from the central type are designated 'varieties'.

Conclusions which he drew from his discussions included:

3 The descendants of varieties frequently revert to the central type of the species from which those varieties originated. But we cannot show that all varieties eventually do thus revert.

4 The effects of cultivation, in rendering varieties more different, and perhaps more permanently different, from their central types, together with the occurrence of hereditary varieties among wild plants, give plausibility to the supposition that varieties do not always revert to the central types of the species from which they originated.

5 A variety (if such there be) in which the tendency to reproduce its own like has superseded the tendency to revert to the central type of the original species, would possess the essential character of a species in itself – namely, its own distinct and permanent central type. It has not yet been proved that any such variety exists; neither can it be disproved.

6 The discordant opinions of botanists, as to which plants are varieties only – the occurrence of varieties intermediate between presumed species – the power of changing from the central type into varieties, and back again to the central type – the tendency of some varieties to become hereditary, probably in obedience to the law of 'like producing like' – with other facts, point towards the conclusion that varieties may gradually become species ...[23]

---

[22] Mayr 1972a, 64 and Mayr 1982, 254–69.
[23] Watson 1845f, 225–8.

The 'essentialist' component of Watson's species concept does not appear to be the holdover of an obsolete metaphysical notion, but rather an attempt to explain findings which now are accounted for with the concepts of genotype, phenotype and ecotype.

If, to modern readers, Watson's ideas seem fairly cautious, that was not the reaction of many readers of the *Phytologist*. William Wilson wrote the editor and complained that in Watson's remarks 'too much has been conceded to the transmutationists'.[24]

George Combe had shown a brief interest in transmutation in 1836 when Watson appended his views on the subject to his rebuttal of William Scott. In January 1846, when he resumed his questions to one disciple, Watson, he was motivated not only by curiosity, but also by desire to write a useful review of *Explanations: A Sequel to Vestiges of the Natural History of Creation* (1845) by another disciple, Chambers. Combe's review, in *The Phrenological Journal*, dwelled on the usual phrenological themes.[25] However, he had asked Watson for some general evidences supporting the possibility of transmutation, and he added Watson's signed reply to the end of his review. Watson referred phrenologists to his series in *The Phytologist* for more details. However, he added for *The Phrenological Journal* an account of the idea of the transmutation of primitive leaves into floral and other organs. This concept goes back to Goethe's *Metamorphose der Pflanzen* (1790), though Watson did not say where he had obtained the idea.[26]

After Watson read *Explanations*, he lamented to Combe (17 June 1846) that its author was 'not really conversant either with botany or invertebrate zoology', and consequently, 'He has the unlucky knack of selecting the questionable, or probably erroneous, where he might get far better evidences.' Most annoying was the praise which *Explanations* wasted (p. 166) on Forbes's paper (1845) on the geographical divisions of the British vegetation: 'The only real basis of Forbes's paper was printed in a small botanical work of mine in 1835. The rest is pure invention, not knowledge.' Watson claimed that he had pondered transmutation for fifteen years, and could write a sounder defense of the idea than any yet published, 'But if I ventured to propound mine, it would meet with violent hostility; and as I feel conscious that it cannot be thoroughly proved by a single individual, the probability is that it will die with me.' The difficulty was that 'each department of science is so large that no single writer can become so far conversant with more than one department as to judge fairly on the relative value of its facts and evidences, in reference to the hypothesis'.

---

[24] Wilson 1846. On William Wilson (1799–1871), see Desmond 1977, 668.
[25] [Combe] 1846; also [Combe] 1845.
[26] [G. Combe] 1846, with Watson's letter of 29 January 1846 on 171–5; the original of this letter is in the Combe correspondence. On Goethe's work, see Eyde 1975 and Wells 1972.

Despite frustrations aroused by reading *Vestiges* and *Explanations*, Watson wanted 'a friendly acquaintanceship' with their author (17 June 1846). This was more than an idle comment, for in January Combe had requested permission to transmit some of Watson's ideas to the anonymous author. However, Combe was undoubtedly sworn to secrecy, and he did not respond to Watson's hint. Later, acting on a hunch or gossip, Watson sent 'a few printed pages', perhaps from his writings in *The Phytologist*, to Robert Chambers, only to receive in return a strenuous denial of authorship (as Watson reported to Combe, 14 May 1847).

Combe sent Watson his pamphlet, *On the Relation between Religion and Science* (1847), and in return, Watson explained (14 May 1847) why he rejected Combe's belief in a benevolent designer of nature. It was no good claiming that the sufferings in nature are for man's education:

> The human race is only one species, among half a million of other organic species now on the earth. A large proportion of the injuries and sufferings (which you view as penalties, designed to instruct when they fall on man) do fall to the lot of the 499,999 other organic species also. Yet, these species benefit not thereby as species. The experience of the individual dies with the individual, and rarely profits more than itself, often profiting itself only to a small degree, or no degree at all.
>
> Other organic species, which inhabited the earth anterior to man, to all appearance must have suffered in the like manner. Human improvement or instruction is too remote to be deemed the purpose of their sufferings.

Watson emphasized what Darwin later called a struggle for existence:

> ... while the species is kept up by some more fortunate or favoured individuals, a vast number of individuals die prematurely; or, from one injurious cause or other, never attain the full growth and health of their species. The lower we go in the scale of development, the more truly is this the case; so that Man, instead of being at present an unfortunate exception, is, in fact, the fortunate or favoured exception.

Animals rob, bully, and murder each other pretty much as men do. Nevertheless, there was some room for optimism:

> Nature has not only created but also destroyed numerous species.
>
> In destroying the earlier species, Nature appears to have quashed them gradually, and gradually introduced fresh species to take their places.
>
> As far as we see, the earlier species were those only of simple organization, and especially those having less endowment of mental qualities.

> Such simply organized and endowed species have continued to be created up to the present state of things. But along with these, there has been a gradual introduction among them of other species more complex in organization, and superior in nervous and cerebral (mental) endowment.
> Apparently the race of Man is one of the latest created species – a species scarce equalled in complexity of physical structure by any other; and also immeasurably beyond any other in mental endowments.
> The grand difference between Man and other organic beings is this. With them, the experience of the individual dies with the individual; being of no value to the species, which remains as it was, neither more nor less capable of adapting itself to external circumstances, whether these change or remain the same. The experience of the individual man can be communicated pretty fully to his fellows, can accumulate with age, and be converted into a sort of experiential capital for the use of the race or species.

Encouraging as this is, man is not exempt from many grim aspects of nature which effect other species:

> Individuals of one species afford food to those of another. A constant struggle is going on between individuals of the same species, in which the stronger triumph, and the weaker fall victims. The latter are killed, or die, or live under more of the suffering and less of the enjoyment of life.
>
> ... The history of mankind shows that individuals of the human species have constantly been sacrificed, or injured, for the advantage of other individuals ... The red and black races are sacrificed to the white; the feebler or inferior whites to the stronger or superior.

He hoped that humanity would use reason and accumulated knowledge to improve the human condition. Some day it should become possible to raise healthy children, 'thus decreasing the inferior specimens of the race, until no very bad ones remain'.
Combe responded (17 May) that 'James Wadsworth, the Proprietor of the largest & most fertile estate in the State of New York, lying on the Genese River, said, "In *The Constitution of Man*, you have developed a new and the true religion."' Neither Watson nor Combe could sway the other on the subjects of design and superintendence in nature, yet Combe wanted to publish part of Watson's lengthy letter. Watson agreed (20 May 1847) but asked that it be attributed to an anonymous friend. Combe did express doubts 'on the question of nature allowing superior individuals and tribes of men benefitting themselves by sacrificing their weaker brethren'. Watson sent him new examples to support his previous claims (27 May),

and he believed that, because of the interdisciplinary aspects of the phenomena encompassed by evolutionary theory – geology and perhaps astronomy, as well as virtually all areas of biology – an understanding of how evolution works could only be achieved by a vast cooperative effort. Yet he was convinced that, because of religious and maybe political implications, society would not tolerate such an endeavor. Hence, the anonymous and mediocre *Vestiges* and *Explanations* would likely hold the field as the most comprehensive presentation. Aside from his speculations which Combe published anonymously,[27] Watson's further published contributions to evolutionary biology were modest and mainly within the context of species-by-species scrutiny of the British Flora.

Darwin's Orientation and Studies

Watson's understanding of evolution during the period 1832–47 was exceeded only by Darwin's. There are striking similarities in their thought, and a crucial difference. Although Darwin had been exposed to evolutionary ideas since reading his grandfather's writings as a teenager,[28] it was only in March, 1837 when John Gould, the ornithologist, began discussing with Darwin whether his Galapagos bird specimens were of new varieties or new species that Darwin became convinced that species evolve.[29] As far as conviction is concerned, Watson had a head start of several years, but in education, breadth of experience and openness of mind, Darwin had an edge. Whatever theoretical progress Watson made between 1834 and 1845 seems to have been confined to his thinking, and not written down. His written investigations were at the empirical level. In contrast, Darwin wrote down both his theoretical ideas and the evidences which he collected from his own studies and from the natural history literature. He began keeping transmutation notebooks in July 1837, he read widely in the natural history literature searching for relevant information, and from what he recorded he wrote progressively longer sketches of his ideas in 1842 and 1844.[30]

Darwin's 1844 essay is somewhat shorter than Chambers's *Vestiges*, but unlike Chambers, Darwin did not aspire to describe the evolution of all of nature; the evolution of biological species was quite broad enough a subject for him, and he

---

[27] Portions of Watson's letter of 14 May 1847 were quoted in Combe 1847, Appendix IX, 494–8 and in Combe 1857, 184–5.
[28] Colp 1986b. Ghiselin 1976.
[29] Sulloway 1982a and 1982b.
[30] For the notebooks, see Darwin 1987 and Sheets-Pyenson 1981b; for the 1842 and 1844 sketches, see Darwin 1909 and Kohn et al. 1982. See also Browne 1995, chaps 15–18, Desmond & Moore 1991, chaps 15–21 and Ospovat 1981.

handled it with more sophistication than anyone else. He finished the draft of his 1844 essay on 5 July and sent it to a copyist,[31] at about the same time Chambers finished the manuscript of *Vestiges*. Darwin received the copy in September, and wrote instructions for its publication should he die. It was already far superior to *Vestiges*, which came out in October, but he felt it was not yet strong enough to overcome the hostility toward evolution among the majority of naturalists. Since Darwin knew where he was headed, and since he could see that *Vestiges* lacked expertise, it did not have the same liberating influence on him that it did on Watson. Instead, the critical reviews of it, especially those published by Adam Sedgwick and T.H. Huxley, made Darwin more cautious in developing his own arguments which could withstand such criticism.[32]

---

[31] Darwin 1959b, 11.
[32] Browne 1995, 457–70. Desmond & Moore 1991, 320–3. Egerton 1970b. Schwartz 1990, 130–40.

Chapter 12

Darwinian Parallels and Contrasts, 1809–58

[My father] was generally in high spirits, and laughed and joked with every one – often with servants – with the utmost freedom; yet he had the art of making every one obey him to the letter.

Darwin, 1959a, 39

What makes Watson a renegade?

Darwin to Joseph Hooker, 27 June 1845

Early Life and Personality

Shrewsbury is about forty miles southwest of Congleton. Like Watson, Charles Robert Darwin was from an upper-middle-class family. Born in 1809, he was five years younger than Watson. Both lost their mothers before leaving home, but Darwin was only eight when he lost his.[1] Neither father remarried; in both families older sisters assumed the mothering functions for the younger children. Charles was a younger son but was born into a family that did not practice primogenture (the Darwins being liberal Whigs). His older brother, Erasmus Alvey Darwin (1804–81), who was born and died in the same years as Watson, was Charles's closest childhood friend, and they remained close as adults. Sister Caroline, Charles eventually decided, had been 'too zealous in trying to improve me', and he would wonder when he entered a room where she was: 'What will she blame me for now?' Charles's response to her admonishments was to '[make] myself dogged so as not to care what she might say', and he expressed no hatred toward her.[2] As a child, he also was fond of the family gardner, John Abberley.

In school, Charles reacted to the classical curriculum just as Hewett did – seeing it as boring and irrelevant. He found alternative interests in chemistry (helping

---

[1] Browne 1995, chap. 1; Darwin 1959, 21–46; Desmond & Moore 1991, chap. 1; Freeman 1978, 66 ff.
[2] Darwin 1959, 22.

brother Erasmus with experiments), collecting beetles (Watson did the same) and hunting birds. Nature hobbies probably served as a sublimation for the teenaged Charles to a similar extent as with Hewett. However, Charles's social interactions with middle-class girls seemed to have been much more extensive and successful than Hewett's.

Nevertheless, Charles also reached adulthood with disabilities, which one physician-psychiatrist-historian tied to a neurosis, but which more recently another physician diagnosed as allergies.[3] He may have repressed the trauma of his mother's death and worried about pleasing an older sister and his father, but he did not have any drawn-out emotional struggle. He was expected to follow in his father's footsteps and become a medical doctor, and went to the University of Edinburgh to do so. Despite doing fairly well in his studies, Charles, like Hewett, abandoned his father's plan for his profession when he realized that his personality was unsuited to the practise of medicine.

Charles loved and respected his father, and he never repudiated his father's secular outlook and liberal politics. When he became an evolutionist, he was merely upholding a family tradition begun by grandfather Erasmus Darwin. Although the Darwins had stronger ties to science than did the Watsons, when Charles abandoned medicine, he temporarily accepted his father's alternative for him: going to Cambridge University and preparing to become a clergyman.

Early Thoughts on the Brain and the Emotions

Darwin attended the University of Edinburgh in 1825–27, and left one academic year before Watson arrived. While there, phrenology became a casual interest, for he also associated with radicals. After he switched to Cambridge University in the autumn of 1827, however, the political influences on him were liberal. This combination of only a weak interest in phrenology at Edinburgh and a shift from radical to liberal influences at Cambridge made him receptive to the anti-phrenological arguments of a respected family friend, Sir James Mackintosh. They dined in Cambridge in 1830, and Darwin's attachment to phrenology was 'entirely battered down ...'[4]

At Cambridge, Darwin developed interests in botany, geology, and the philosophy of science.[5] He did not feel compelled to seek immediately an alternative

---

[3] Colp 1977 and 1984. Smith 1990.
[4] Darwin to W. Fox, 3 January 1830; Darwin 1985–, I, 96–7. Desmond & Moore 1991, 23, 33, 43 and 77.
[5] Desmond & Moore 1991, chaps 4–7. Ruse 1979, 30–74.

view on mind and brain to that of phrenology. He was free to reopen the subject of mind and brain from an evolutionary perspective in July 1838.[6]

Early Studies on Biogeography

Darwin's interests in natural history found an outlet at Edinburgh in a fascination with marine invertebrates, which he collected with a zoologist, Robert Grant. At Cambridge, he acquired a casual acquaintance with botany as a result of his friendship with Botany Professor John Stevens Henslow (1796–1861). During his last year at Cambridge Darwin read Humboldt's travel narrative and became excited about the adventure of natural history exploration.[7]

Henslow helped arrange for Darwin to take a voyage around the world as a naturalist on *HMS Beagle* (December 1831–October 1836). During those years, his reading of Lyell's *Principles of Geology* and his collecting of animals, plants and fossils aroused his curiosity about the causes of species' distributional patterns. On the voyage, Darwin sometimes helped Captain Robert FitzRoy make geographical measurements. Fitzroy had fine instruments, and his meticulous attention to measurements helped establish the science of meteorology; Darawin was impressed by Fitzroy's careful collection of data.[8]

After returning to London, Darwin distributed his various collections of specimens to experts for study. He also studied his notes and specimens and produced a travel book which was a worthy successor to Humboldt's, both in interest and in scientific importance: *Journal of Researches into the Geology and Natural History of the Various Countries Visited by H.M.S. Beagle, under the Command of Captain Fitzroy, R.N. from 1832 to 1836* (1839). As in Humboldt's *Personal Narrative*, biogeography appears in intermittent accounts scattered along the way, but the book was so successful that most contemporaries with an interest in the subject read it and absorbed Darwin's contributions.[9]

---

[6] Darwin 1987, 517–96. Browne 1985 and 1995, 382–4, 425 *et passim*. Burkhardt 1985. Gruber & Barrett 1974. Richards 1987, chaps 2–5.

[7] Allan 1977, chaps. 1–3. Ashworth 1935. Barlow 1967, 1–19. Darwin 1959a, 44–68. Desmond & Moore 1991, chaps 3–6. On Henslow, see Desmond 1977, 303; Stafleu & Cowan 1976–92, II, 163–4.

[8] Basalla 1972; Browne 1995, 179–80 *et passim*; Burton 1986; Mellersh 1968.

[9] Allan 1977, chaps. 4–6. Barlow 1967, 25–147. Browne 1983, 124–6; 1995, 133–5, 176–7, 415–7 *et passim*. Darwin 1839 and 1959a, 67–104. Desmond & Moore 1991, chaps 7–19. Egerton 1970a. Richardson 1981.

Private and Family Life

Darwin married his cousin, Emma Wedgewood, in 1839, and they lived in London until 1842. However, the air pollution was bad for his health, and he preferred the countryside to the congestion and noise of the city. He also wanted more privacy than was easily available there. They bought their permanent house at Downe, a village south of London and about twenty miles southeast of Watson's home at Thames Ditton. Darwin got along well with his many relatives, some of whom lived in London, and he might have been more gregarious than Watson, except that his chronic health problems frequently inhibited him.[10]

Darwin established a home life to suit his needs. He and Watson developed rather similar domestic lives, except that Darwin had ten children, seven of whom survived into adulthood. Their respective retreats to the countryside just south of London allowed both Watson and Darwin to have access to the resources a scientist needed, but without having to deal with the urban environment very often.

Outlook and Social Responsibilities

Charles's paternal grandfather, his father, and older brother were all non-religious theists. His Wedgwood grandparents and his mother were Unitarians. But after his mother died his sisters raised him in the Church of England, and the first cousin whom he married in 1839, Emma Wedgwood, was also Anglican.[11] Charles studied for the Anglican ministry at Cambridge University, primarily because that profession would have allowed him time to pursue natural history studies. He never became a clergyman, because of the diversion of the voyage of the *Beagle*. An early association with Robert Grant and other free thinkers at Edinburgh, the anti-fundamentalism in Lyell's *Principles of Geology*, and Darwin's own theory of evolution by natural selection were undoubtedly powerful influences. He did not entirely give up on Christianity, however, until he spent a week watching his daughter, Annie, die at age ten, in 1851. (If Darwin suffered from hereditary allergies, as Fabienne Smith thinks, Annie's fatal illness may have involved the same malady.[12]) Charles then became an

---

[10] Allan 1977, chaps 5–7. Browne 1995, chaps 15–21. Darwin 1959a, 82–115. Desmond & Moore 1991, chaps 14–20. Stone 1980, 451–728.

[11] Browne 1995 is helpful on the religion of these relatives; see their names in her index.

[12] Smith 1990. Browne 1995, 498–501, cites the traditional diagnosis of 'bilious fever' and does not refer to Smith's article.

uncertain theist, and remained so for the rest of his life.[13] Since childhood, natural history had meant more to him than religion, and when he gave up on Christianity, a scientific outlook was already part of his thinking.

Charles's grandfather Erasmus had been a radical freethinker in politics as well as in religion, but his father, Robert Waring Darwin, was liberal, not radical. Therefore, Charles's liberal outlook was a family tradition, not a generational rebellion. He also lived on inherited wealth, and was not inclined to drift into radicalism. His Edinburgh associate, Robert Grant, was a radical Lamarckian, but instead of being influenced by him, Darwin drifted away from Grant.

Charles Darwin's level of involvement in politics and in society was about the same as Watson's. He had an average middle-class interest in politics, which he discussed with friends and relatives, but he never went as far as Watson did by publishing a political pamphlet. He seems to have been slightly more involved in local charitable work than Watson was, giving his time as well as his money. Since Emma was active in the Church, Darwin found it convenient to work personally with the local minister on charitable matters.

Darwin and the Azores

While Watson had spent slightly more than four months in the Azores, the Beagle made its last stop there for only a few days, arriving on 19 September 1836, and leaving for England on the 25th. Darwin explored Terceira, one of the larger islands which Watson regretted not being able to visit. Since FitzRoy had established its geographical position in 1836, Vidal found it unnecessary to do more in 1842. The arable lands were so heavily cultivated, and everything was so parched by September, that there was little to pique Darwin's interest. The few species of European birds and insects which he saw did not astonish him. Only the people and the volcanic geology interested him this late in his voyage. In contrast to Watson's negative evaluation of the Azoreans, Darwin thought: 'It was pleasant to meet the peasantry; I do not recollect ever having beheld a set of handsomer young men, with more goodhumoured expressions.'[14] By the time Watson published his findings, however, Darwin was no longer an explorer, but a theoretician in quest of data. He would become Watson's most appreciative reader.

---

[13] Brooke 1985. Brown 1986. Browne 1995, 325–7, 396–9, 411, 438–9 and 513. Cornell 1987. Darwin 1959a, 85–96; Darwin 1987, 517–641. Desmond & Moore 1991, index, 780–1. Gillespie 1979. Moore 1988b. Ospovat 1981, *passim*.

[14] Darwin 1839, 596.

Involvements with Science and Scientists

There are interesting similarities and differences between Watson's and Darwin's professional involvements during the 1840s and 1850s. Both men were hard-working scientists and persistent letter-writers. Darwin, indeed, might be the most prolific letter-writer in the history of science.[15] Even in the 1840s and 1850s he achieved an impressive volume of correspondence with a fairly large number of scientists: his published correspondence for these decades runs to five-and-a-half lengthy volumes.[16] Since both Watson and Darwin seem equally committed to science and to have invested about the same amounts of time in it, if we ask why Darwin was more productive of correspondence during these two decades, the answer seems to be because of his broader scientific interests and his more agreeable personality. Within botany, Watson's interests were fairly broad, including systematics, plant geography and evolution; his interests in plant geography also ensured some interest in environmental factors and physiology. Yet in 1840 when Watson severed his scientific commitment to phrenology, he never became involved with any new subject; he resumed his earlier botanical activities. On the other hand, Darwin's commitment to evolution shortly after his return to England in 1836 had the effect of expanding his already broad interests in zoology, geology, paleontology, and biogeography. It helped him develop an interest in botany that his best friend, Joseph Dalton Hooker, was a botanist, as was his mentor of the 1830s, John Stevens Henslow.[17]

Both Darwin and Watson were Fellows of The Linnean Society of London, and since it was tilted toward botany, one might expect Watson to have been the more active of the two in its meetings and affairs. That, however, does not seem to have been the case, in part because Watson was active in the Botanical Society of London, which met his needs better than did The Linnean Society. Darwin was also elected a Fellow of the Geological Society of London and the Royal Society of London – both being élite scientific organizations which never included Watson. Both men attended a very few annual meetings of the British Association for the Advancement of Science – an inclusive organization whose meetings Darwin would probably have attended more often had his health been better. Watson had two unpleasant experiences at BAAS meetings which probably inhibited his interest in the organization: its rejection in 1838 of a phrenology section at its meetings, and his conflict with Forbes in 1846.

---

[15] The magnitude and diversity of his correspondence is nicely indicated by Moore 1985.
[16] Darwin 1985–; see half of II and III–VII.
[17] Allan 1977, chaps 3–10. Barlow 1967. Desmond & Moore 1991, chaps 20–32. Charles Darwin and his brother Erasmus were lifelong friends with much in common, except that Erasmus was not a practicing scientist, and therefore he became less involved in Charles's life than Hooker was.

Both Watson and Darwin were liberals with radical tendencies. Watson created an unfavorable impression when he admitted in his applications for botanical chairs that he supported phrenology and did not adhere to the beliefs of the Church of England. Although Darwin was never in a similar situation, if he had been, it is unimaginable that he would have exposed his radical ideas voluntarily. Whereas Watson placed a high value in asserting his individuality, Darwin placed a high value in preserving harmony. However, this difference did not mean that Watson was more open and Darwin more secretive. On the contrary, Watson was inconsistent, being sometimes open and sometimes not. Darwin was open to the extent which he judged prudent and silent beyond that. Considerateness and lack of deviousness were major assets in his professional and personal relationships.

Darwin's considerateness developed early in his life. It might have been reinforced professionally by the disappointment of having Louis Agassiz challenge Darwin's explanation of the parallel roads of Glen Roy in Scotland. On the basis of Darwin's experience with uplift during earthquakes in the mountains of Chile, he had made a careful study of the Glen Roy formations and argued in 1839 that they could be explained by the same phenomenon – uplifted beaches. The next year, Agassiz studied them and realized that they were caused by glaciation. In 1848 Darwin still hoped to salvage his own interpretation, but in 1861 he admitted that the glacial interpretation seemed conclusive.[18]

Darwin and Joseph Hooker met briefly in 1839 before Hooker left on a four-year voyage into the Southern Hemisphere. At the time, Hooker was reading Lyell's copy of the proofs for Darwin's *Journal of Researches* (1839) about his voyage on the *Beagle*, but did not say so. Hooker was eight years younger than Darwin, and there was no clue at the time that they would later become close friends. However, in November 1843, when Hooker had only been home from his voyage a short time, Darwin wrote to bring to his attention Darwin's own plant specimens from his voyage on the Beagle; Henslow had agreed to publish accounts of them, but had not. They exchanged only a few letters before Darwin confided to him his belief in the transmutation of species.[19] Since Darwin was cautious about his radical ideas, this was a surprising confession, indicating rapid gain of mutual trust. He did not win over Hooker to a belief in transmutation until after Hooker read the published version of *On the Origin of Species*, in the spring of 1860 – even then he claimed to

---

[18] 'I sh[oul]d have been more sorry to have been proved wrong on it, than upon almost any other subject.' Darwin to Robert Chambers [June 1848]; Darwin 1985–, IV, 148. Darwin conceded the glacial interpretation on 6 September 1861: Darwin 1985–, IX, 255–7; cf. also 429–59. Darwin's 1839 article is reprinted in Darwin 1977, I, 87–137. Agassiz 1840; Barrett 1973; Rudwick 1974.

[19] Darwin's confession letter was written 11 January 1844. All the letters in this exchange are in Darwin 1985–, II, between 408 and 422, and III, 1–3.

Figure 12 Charles Darwin (1849), by T.H. Maguire.

have been won over by his own botanical studies rather than by Darwin's book[20] – but if Hooker was his own man, he nevertheless did not dismiss Darwin as a superficial thinker in 1844, as Darwin feared he might.

He accepted Darwin's invitation to study the Galapagos plants and compare them to floras of other islands and South America. Hooker finished the taxonomic part of his study in 1845 and presented his results before the Linnean Society at three different meetings. The second part, on biogeography, he discussed with Darwin in a letter of 25 March 1846 and then sent him a draft in November, to which Darwin responded enthusiastically: 'it is without comparison, the best essay on geograph. distrib. in any class, which I have ever met with ...'[21] Darwin did follow this complement with suggested improvements, and Hooker presented the revised version at two meetings of The Linnean Society in December.[22] By then they had established a close friendship that lasted a lifetime and included numerous scientific collaborations.[23]

Darwin may have met Forbes at the annual meeting of the British Association in August 1839 at Birmingham, but Darwin does not seem to have been among the group of young naturalists whom the ebullient Forbes gathered for dinners at the Red Lion. Those gatherings became annual events, and the group took the name Red Lions. Neither Darwin, Watson nor Hooker seem ever to have joined.[24] Darwin probably got to know Forbes mainly after he obtained the Professorship of Botany at King's College, London in 1842, as they were then both active in the Geological Society of London. Darwin was present at GSL's meeting on 14 December when Forbes was elected Curator.[25]

Forbes established a good reputation as a invertebrate zoologist and paleontologist, and he won Darwin and Hooker's friendships by a combination of friendliness and competency. Darwin had him name and describe eleven of his *Beagle* fossil shell species.[26] Besides a technical competency in describing species, Forbes had an interest in explaining the distributions of species. He was well-read and industrious, but less cautious than Darwin, Watson, Hooker or Lyell in drawing theoretical conclusions. Darwin, not being a botanist, was a silent observer of the Watson–Forbes conflict and undoubtedly accepted Hooker's unfavorable judgment of Forbes's botany. However, there were also zoological, geological and geographical

---

[20] Huxley 1918, I, 516–20; Turrill 1963, 76.
[21] Darwin to Hooker [23 November 1846] in Darwin 1985–, III, 369. Hooker's letter on 25 March 1846 is in III, 303–6.
[22] Hooker 1851b; read on 1 and 15 December 1846; partly reprinted in Hooker 1953, 121–9, followed by Turrill's commentary, 129–33. See also Porter 1980.
[23] Browne 1978.
[24] Gardiner 1993. Huxley 1918, I, 370, n. 1. Rehbock 1983, 184.
[25] Darwin 1985–, II, 340, n. 4.
[26] Forbes 1846b.

dimensions of Forbes's ideas which Darwin could judge for himself. His own handling of disagreements with Forbes provides a telling contrast to Watson's.

In June, 1845 – before Forbes gave his talk 'On the Distribution of Endemic Plants' to the BAAS – Darwin heard about it and expected it to be 'very good work'. Darwin even felt that Forbes might forestall him in discussing 'the relation between the present alpine & Arctic floras, with connection to the last change of climate from Arctic to temperate, when the then Arctic lowland plants must have been driven up the mountains'.[27] However, on 3 September, when he understood Forbes's arguments, he immediately parted company with him on geological grounds. To Hooker's initially favorable reaction, Darwin responded:

I am very glad that Forbes has one such thorough appreciator as you: I fancied I was bold enough in upheaving & letting down our mother earth, but Forbes beats me hollow, without any proof to speculate on Ireland & Portugal having been once connected ... this boldness is undoubtedly the direct consequence of Lyell's *Principles*.[28]

Hooker defended Forbes, and Darwin stood his ground.[29] Before long, Darwin expressed his doubts to Forbes, who responded with a letter and geological map (25 February 1846) that went beyond his talk of the previous summer. Darwin remained unconvinced. To both Forbes and Hooker, Darwin confided that his own knowledge of the geology in question might be inadequate, but he still felt that Forbes's claims rested on 'too great [a] hypothetical basis'.[30] Hooker was curious to see Forbes's letter and map, and after studying them began to have his own doubts (2 March 1846), though he did not reject Forbes's botanical documentation until he had read the full published version.[31] Darwin was pleased to hear Hooker's doubts and optimistically imagined that they agreed on Forbes's deficiencies.[32] In reality, however, Darwin was rejecting Forbes's geology, and Hooker was rejecting his

---

[27] Darwin to Hooker, [27 June 1845]; Darwin 1985–, III, 207.

[28] Darwin to Hooker [3 September 1845]; Darwin 1985–, III, 250. His blaming the influence of Lyell's *Principles* on Forbes seems accurate. Lyell complained to Forbes about borrowing his ideas without acknowledgment and Forbes apologized. Rehbock 1983, 185.

[29] Darwin to Hooker [10 September 1845]; Darwin 1985–, III, 252–3.

[30] Darwin to Hooker [25 February 1846]; Darwin 1985–, III, 294. Forbes's letter to Darwin [25 February 1846], with geological map, is on 290–3. Darwin's letters to Forbes have not survived.

[31] Darwin sent Forbes's letter and map to Hooker some days before Hooker responded on 2 March 1846; Darwin 1985–, III, 294–5. Hooker's rejection of Forbes's botanical data and conclusions on 28 September 1846 is quoted above, pages 125–6.

[32] Darwin to Hooker [13 March 1846]; Darwin 1985–, III, 300.

botany. Hooker did not share Darwin's aversion to inventing land bridges to account for discontinuous distributions of species and was only won over to Darwin's perspective in 1867.[33]

The land bridge controversy was a typical one for science, in that each side had a part of the truth. Sometimes former land bridges have disappeared, as between Britain and France or between Siberia and Alaska. Today the disappearance of many of these is explained by the rise of sea level after the ice ages rather than by the subsidence of the sea floor. Although the previous existence of ice ages was recognized by the mid-1840s, their impact on sea level was not. Another important unknown factor was the lateral instability of the earth's crust, explained since the 1950s by the theories of continental drift and plate tectonics. Land masses which were once close are often now much further apart. Darwin was right to believe that coconuts can drift on ocean currents to distant shores, but that idea could never explain the discontinuous distributions of monkeys between Africa and South America. One of Darwin's responses to this disagreement between what Rehbock calls the 'migrationists' and the 'extensionists'[34] was to conduct a series of experiments on the viability of seeds which had floated for different periods of time in sea water. Whether coincidentally or not, he only began these experiments after Forbes died.[35]

Although Darwin's seed viability experiments may have been delayed more by his work on barnacles than by his concern for Forbes, the same cannot be said for his delay in contacting Watson. Hooker had sent Watson's studies on the plant geography of the Azores to Darwin on 12 December 1844, and a few months later, after Watson began publishing his responses to Robert Chambers's anonymous book, *Vestiges of the Natural History of Creation*, Hooker referred to Watson as 'a renegade'. Darwin wanted to know why, and Hooker responded because of Watson's belief in transmutation. In view of what Darwin had confided in Hooker in January 1844, that response must have seemed surprising. Darwin replied: 'You will be ten times hereafter more horrified at me, than at H. Watson.'[36] Darwin was busy revising his *Journal of Researches* for a second edition, and he seems not to have pursued his curiosity about Watson. In September 1846, Darwin made no effort to find Watson,

---

[33] Hooker 1867. Rehbock 1983, 190.

[34] Rehbock 1983, 187.

[35] Forbes died 18 November 1854; Darwin published the first of his papers on seeds floating in sea water on 14 April 1855; the whole series of articles is reprinted in Darwin 1977, I, 255–63. Completion of his barnacle studies in 1854 gave Darwin time for his seed studies.

[36] Quotation, Darwin to Hooker [11–12 July 1845]; Darwin 1985– , III, 217. When Hooker first called Watson a renegade seems unknown, but Darwin's query about it was 27 June 1845; ibid., 207. Hooker's letter to Darwin explaining what he meant was dated 5 July 1845; ibid., 211.

but would have been curious to meet him while they were both at the BAAS meeting. However, it was at this meeting that Watson refused to back down on his plans to attack Forbes in print, therefore Darwin waited until the year after Forbes died before he finally wrote to Watson.[37] Their exchange of letters was very helpful to Darwin when he wrote *On the Origin of Species*.

Darwin seems never to have gratutiously attacked another scientist, but he did have to consider how to respond to their attacks on him. Perhaps there were no direct attacks until he published *On the Origin of Species* (1859). Then he had two kinds of responses to published attacks: first, he complained privately to Hooker, Huxley, and to other friends and colleagues when he felt that criticisms were unfair, and secondly, he attempted to respond in print to significant scientific criticisms. Later editions of the *Origin* and other writings gave his courteous responses to critics without any of the petulence which he sometimes expressed privately and which Watson commonly expressed in print.[38]

In the first edition of the *Origin*, which was only lightly documented, Darwin acknowledged Watson's assistance on eight different pages (six of which are indexed under Watson's name). On the first such page, he referred to 'Mr. H. C. Watson, to whom I lie under deep obligation for assistance of all kinds ...'[39] Although Darwin was sincerely grateful for Watson's help, he also wanted to be sure he did not risk neglecting adeqeuately acknowledging Watson's help. Watson was pleased with the acknowledgments, and wrote to Darwin on 21 November: 'Once commenced to read the *Origin* I could not rest till I had galloped through the whole.' Watson was stingy with compliments, and never gave them without some sort of qualification, but this letter undoubtedly contained the greatest compliment which he ever paid to anyone: 'You are the greatest Revolutionist in natural history of this century, if not of all centuries.'[40] Darwin was to receive many fine compliments, but none finer than that. Watson wrote his letter a few days before the book went on sale, and he probably read a copy given him by Darwin.[41]

Usually when Darwin lost the esteem of friends or colleagues it was because evolution was an unpopular theory. However, there were a few awkward episodes of other sorts with which he had to contend. The most famous one arose when he received

---

[37] Only Watson's replies survive; the earliest is dated 11 July [1855]. The two letters from Watson which the editors of Darwin's *Correspondence* guessed were written in November [1854] should be dated November [1855].

[38] Hull 1973; Peckham 1959; Vorzimmer 1970.

[39] Darwin 1859, 48; this and 53 were missed in Darwin's index; Watson was also cited on 58, 140, 176, 363, 367 and 376.

[40] Watson to Darwin, 21 November [1859]; Darwin 1985–, VII, 385.

[41] John Murray, his publisher, sent Darwin a copy on 2 November and it went on sale 25 or 26 November; Desmond & Moore 1991, 476.

in the mail Alfred Russel Wallace's unsolicited paper on evolution by natural selection in 1858 while Darwin was writing his big book on the subject. Since their ideas were rather similar, Darwin appealed to Hooker and Lyell for advice. Fortunately, Hooker, Lyell and Asa Gray were all familiar with some of Darwin's unpublished writings and could vouch for his having developed his ideas before receiving Wallace's paper. Although later a few poorly informed critics have argued that Darwin did steal from Wallace, the vast amounts of early Darwin manuscripts and letters leave no room for reasonable doubt on the matter. Amusingly, Richard Owen was ambivalent about evolution before publication of the *Origin*, but having become convinced from reading the book that the idea was valid, he then claimed that Darwin was indebted to his own writings beyond the generous extent to which Darwin cited him. Similarly, after Darwin published his theory of pangenesis in 1868, Owen argued that Darwin was wrong in claiming that his ideas differed from Owen's. Other scientists and historians have sided with Darwin in both instances.[42] Also amusing is the fact that both Darwin and Herbert Spencer agreed that Darwin's pangenesis differed from Spencer's earlier published notions on heredity, whereas to modern judgments, their ideas seem remarkably alike.[43] Very annoying was the writer Samuel Butler's claim (1879) that Darwin had plagiarized his ideas from his own grandfather, Erasmus Darwin. This despite the facts that Erasmus Darwin had not thought of natural selection and that Charles Darwin had mentioned him in 'An Historical Sketch' which he added to the third (1861) and later editions of the *Origin*. In fairness to Butler, Charles did make use of Erasmus's ideas of sexual selection and the inheritance of acquired traits. As in the case of Wallace in 1858, Charles Darwin consulted with friends and colleagues about Butler's attacks. He published an appreciative essay on his grandfather (1879), but made no direct reply to Butler.[44] A modern critic has imagined that Charles Darwin made more use of the ideas of Edward Blyth than Darwin acknowledged in the first edition of the *Origin*, though he has convinced few, if any qualified scholars.[45]

This list of awkward situations shows that even a conscientious scientist who is very productive and who ranges widely in his readings and researches can run into trouble. Darwin, in his later botanical writings, was to cite Watson's works often, though he forgot to include him in 'An Historical Sketch' (1861). Watson, however, had already been treated so well in the main text that he did not complain to Darwin about it.

---

[42] Bowler 1990, 162–3. Desmond & Moore 1991, see under Richard Owen in index; Geison 1969, 396–8; Hull 1973, 171–215. See also Desmond 1985.

[43] Geison 1969, 398–404.

[44] Bowler 1990, 172–4. Desmond & Moore 1991, 635 and 640–1. Freeman 1978, 45–6. Ghiselin 1973a and 1976.

[45] Darwin 1859 cited Blyth on 18, 163 and 253. Eiseley 1979 (posthumous) reprints two earlier accusatory articles, three of Blyth's own articles, and an obituary of Blyth. In refutation, see Schwartz 1974 and 1980.

Chapter 13

Stonecutter for Darwin's Edifice, 1847–59

> You seem to expect something from the concluding volume of Cybele, & will thus be disappointed. It will simply be like a block of stone, cut ... [for] an edifice ...
> Watson to Darwin, 3 January 1858

The year 1859 saw a climax to the scientific publications of both Watson and Darwin. In May, Watson published the fourth volume of *Cybele Britannica*, containing interpretations of phytogeographical data on individual species which he had published in the earlier volumes (1847–52). In November Darwin published *On the Origin of Species*, culminating researches he had carried on since 1837. There was some similarity in their goals and strategies, since both wanted to explain the origin of species and both appreciated the relevance of plant geography to the task. Nevertheless, there was hardly any similarity in the organization of these two volumes. An unforeseen circumstance – Darwin's receipt of Alfred Russel Wallace's letter in 1858 – led Darwin to make a readable abridgement of the monograph on species and natural selection which he had begun in 1856, resulting in one of the great classics in the history of science. Watson was not to be deflected by a somewhat different interruption – Darwin's inquiries about British species and genera of plants – and his fourth volume was an energetic attempt to glean insights from all the data he had published.

Barnacle researches and publications had absorbed much of Darwin's attention from July 49 until September 1854, when he began sorting his notes for his species book.[1] Watson was busy simultaneously with his volume IV.[2] It is interesting to compare how effectively the two men used Watson's findings in building their respective syntheses.

---

[1] Darwin 1959b, 12–13.
[2] Watson 1847–59, IV, 3.

A Colleague for Darwin

In December 1843, Joseph Hooker had agreed to study Darwin's plant specimens from the Galapagos Islands. By 11 January 1844, Darwin felt confident enough about their relationship to confess his belief in the transmutation of species, a belief Hooker would not share until after he had read *On the Origin of Species*. Despite this difference in belief – or maybe because of it – they never tired of discussing island floras. Darwin's questions on 23 February began with the Galapagos species in relation to the species on the South American mainland, and led to questions about the relationships between other islands and mainlands, including: 'How is it with the Azores: to be sure the heavy West: gales wd tend to diffuse the same species over that group.'[3] This speculative inquiry was based upon a little experience, as the Beagle's last foreign stop in 1836 had been for almost a week at Terceira, an island Watson never visited.[4]

Since both Darwin and Hooker had several research projects going simultaneously, their discussion of island floras did not proceed rapidly. On 12 December, Hooker sent for Darwin's perusal issues of the *London Journal of Botany* containing Watson's articles on the Azores. Darwin later obtained for himself those issues, and his annotated copy of Watson's accounts survives. On Christmas Day he wrote to Hooker:

> Watson's Paper on Azores has surprised me much; do you not think it odd, the fewness of peculiar species, & their rarity on the alpine heights: I wish he had tabulated his results: c^d. you not suggest to him to draw up a paper of such results, comparing these isl^d with Madeira; surely does not Madeira abound with peculiar forms? A discussion on the relations of the Floras, especially the alpine ones, of Azores, Madeira & Canary Is^d would be, I sh^d think, of general interest: – How curious the several doubtful species, which are referred to by Watson, at the end of his Paper; just as happens with birds at the Galapagos.[5]

Hooker agreed that 'The paucity of peculiar Azorean species is very strange & more particularly the want of W[est] Ind[ies] or N[orth] Am[erican] forms, though the current washes up canoes (if all [reports] be true) on their shores.'[6] Hooker assured Darwin he had written to Watson on the questions raised, and Watson would

---

[3] Darwin 1985–, III, 11.
[4] Darwin 1839, 594–8. His experiences there are discussed above, page 167.
[5] Darwin 1985–, III, 100. Darwin's copies are in the Darwin Collection, University of Cambridge. For some indication of when he first read and then purchased them, see Vorzimmer 1977, 132.
[6] J. Hooker to Darwin, 30 December 1844, in Darwin 1985–, III, 102.

certainly have responded, though he never expanded his studies to include Madeira and the Canary Islands.

In the same letter Hooker teased Darwin about a bedfellow in the transmutationist camp: 'I have been delighted with *Vestiges*, from the multiplicity of parts he brings together, though I do [not] agree with his conclusions at all. He must be a funny fellow.'[7] Darwin had also read Robert Chambers's anonymous book, but he was 'somewhat less amused at it, than you appear to have been: the writing & arrangement are certainly admirable, but his geology strikes me as bad, & his zoology far worse'.[8] When Watson's review of *Vestiges* and his subsequent essays on the origin of species appeared in the *Phytologist* in 1845, Hooker seems to have mentioned them to Darwin, but in such a way as to inhibit rather than awaken curiosity: he claimed that Watson's ideas on progressive development were similar to those in *Vestiges*.[9]

That uninteresting image was reversed when Watson attacked Forbes. Hooker's expression of consternation to Darwin (before 3 September 1846) shows the highest regard for Watson as botanist while lamenting his severe response to what Hooker saw as a fairly minor transgression. This conflict inhibited Darwin, a friend of Forbes, from consulting Watson directly, but Hooker, being friendly to both Darwin and Watson, was willing to serve as an intermediary. Darwin wrote to Hooker on 7 April 1847 and requested information on 'cases of varieties between two other varieties being rare ...'.[10] Hooker posed the question to Watson, and sent to Darwin Watson's very impressive response. Darwin had a copy made of Watson's letter, which copy he then annotated and drew upon when writing his long manuscript in 1856–58 that was eventually condensed into *On the Origin of Species* (1859).[11] His appetite for Watson's expertise having been whetted, Darwin was pleased in June 1847 when Hooker lent him the first two volumes of the *Phytologist* and the first volume of *Cybele Britannica*.[12]

---

[7] P. 102.

[8] Darwin to Hooker, [7 January 1845]; Darwin 1985–, III, 108. For more on Darwin's reactions to *Vestiges*, see above, page 162.

[9] Hooker to Darwin, 5 July 1845; Darwin 1985–, III, 211.

[10] Darwin 1985–, IV, 30.

[11] Watson to J. Hooker, 12 April 1847, is quoted in full, with Darwin's annotations, in Darwin 1985–, IV, 31–2. In his Natural Selection MS, 1856–58, Darwin wrote: 'I applied to Mr. H.C. Watson & to Dr. Asa Gray for their opinions on this head; as from their critical knowledge of the floras of Great Britain & the United States, everyone would place great confidence in their judgment. Both these botanists concur in this opinion, & Mr. Watson has given me a list of twelve nearly intermediate varieties found in Britain which are rarer than the forms, which they connect.' Darwin 1975, 268.

[12] Darwin to Hooker, [10 June 1847], Darwin 1985–, IV, 47 and n.6 on 48. See also Vorzimmer 1977, 138.

For seven years, poor health, the study of barnacles and a variety of other researches diverted Darwin from the questions on species he associated with Watson; in September 1854 he finished work on his four volumes of barnacle studies and began sorting his notes for writing his book on species theory.[13] Forbes died a month later, and when Darwin wanted to consult Watson again, he no longer needed Hooker as intermediary. In 1855 he contacted Watson, who had wanted to meet him for a decade.[14] Coincidentally, Darwin was also shifting from mostly using animals for his investigations to mostly using plants, for plants seemed more convenient and manipulating them more humane. Questions on the spread of species to islands motivated his experiments in 1855 on how long seeds could float in sea water and remain viable.[15] He explained to Hooker on 27 May that these experiments were meant to show 'the *possibility* in the long course of ages of a few plants being transported by currents'. If he got positive results, then: 'the real interesting thing would be to get a list of the Azores plants, & try & get the seeds of as many as I could, & test them ...'. He had either mislaid or not yet obtained his own copy of Watson's study on the Azores, for he asked Hooker: 'Can you give me reference to Watson's Paper or best list of the plants of this archipelago?'[16] By 2 July he decided that he need not test the seeds of all Azores plant species in sea water, but only seeds of species that might have been transported from Europe.[17] Darwin's interest in Azores species was strong enough for him to write on 7 May to Thomas Carew Hunt, the British Consul who had supplied Watson with plant specimens until 1848. Darwin was unaware of Hunt's departure from the Azores, but the letter was forwarded to him in France and he answered Darwin's questions as best he could.[18]

Having compiled a list of the European species in the Azores from several sources, Darwin sent a list of 77 non-domesticated species to Watson with a request for his estimate of how many of them might have drifted there on sea currents.

---

[13] Darwin 1959b, 11–13.

[14] On Watson's early desire to meet Darwin, see his letter to J.Hooker, 6 June [1845]. The editors of Darwin's correspondence guessed that Watson's first two letters, dated 19 and 20 November, were written in 1854. This implies that Darwin wrote to Watson before he knew of Forbes's death on 18 November 1854 – an unlikely situation. Darwin 1985–, V, 239, n. 1. Vorzimmer 1977, 147, shows that Darwin read in *Phytologist* vol. 3 on 27 April 1855. Darwin read Watson 1835a on 15 June 1855; ibid., 150.

[15] Darwin published six notes on this subject, 15 April–29 December 1855, in the *Gardners' Chronicle and Agricultural Gazette* and also a longer article in the *Journal of the Linnean Society* (1857 [1856]) – all of which are reprinted in Darwin 1977, I, 255–8, 260–73.

[16] Watson to Hooker, 27 May [1855], Darwin 1985–, V, 339. On 23 June, Darwin wrote to Hooker that he had re-read (first reading in December 1844) Watson's paper on plants of the Azores while at the Linnean Society in London. Ibid., 357.

[17] Darwin to Henslow, 2 July [1855], Darwin 1985–, V, 365.

[18] Hunt to Darwin, 2 July 1855, Darwin 1985–, V, 366–7.

Watson identified 22 species, which he marked on Darwin's list, as plausibly arriving by sea. Darwin then wrote to both Henslow – to ask how many of these seeds could be collected by his daughters – and to Hooker – to ask which dealers might supply seeds. Darwin also wrote to Hooker, saying that Watson had been most kind, and that Darwin had accepted his invitation to come for a visit![19]

Darwin went in early August for their only face-to-face encounter. Watson undoubtedly showed off the garden in which he grew species and varieties of questionable status. He also showed Darwin de Candolle's *Géographie botanique raisonnée*; and when Watson wrote to de Candolle on 28 October 1855, he reported that Darwin had obtained his own copy and was pleased with it. Much sooner than that, however, Darwin confided his impressions of Watson to Hooker: 'My morning with H.C.W. passed off very prosperously & I had much very interesting talk; but he is rather too sarcastic to my taste. He strikes me as a very clever man.'[20] By October Darwin began to worry about whether he might have a conflict with Watson á la Forbes, and decided to put his cards on the table. Watson responded:

In reference to a remark of your letter, closely following the name of E. Forbes, I claim no property in what I may write during such a correspondence as ours. I simply comment on your ideas, & you are fully at liberty to use or apply such comments in any way you like. You probably deem me touchy or tenacious of aught that has been my own, because I fell foul of Forbes in regard to certain geographical groupings of British plants. – Forbes did not re-examine these groupings, – ascertain their correctness, – & then apply them to his further object. Had he done so, they would have become his also; & I should have said nothing (or little) about his mere omission to state that such groupings were not *original with him*. But he practiced a *fraud* on the British Association & general public of science, by giving on his sole personal authority the results of long investigations & comparisons which he had neither made, nor repeated, nor imitated, – simply misappropriated, & without clearly understanding them.[21]

Even if Darwin felt that Watson had been too severe with Forbes, he was reassured enough to continue their correspondence, and invited Watson to visit Down House on 22 April 1856, along with the Joseph Hookers, Huxley and Thomas Vernon Wollaston. Watson sent his regrets. He might have gone had he been invited alone, but he would not have enjoyed having to compete with other guests for

---

[19] Watson to Darwin, 11 July [1855] and Darwin to Henslow, 14 July [1855], Darwin 1985–, V, 374–6. Darwin to Hooker, 14 [July 1855], Darwin 1985–, V, 377.
[20] 10 August [1855], Darwin 1985–, 403.
[21] Watson to Darwin, 11 October 1855; Darwin 1985–, V, 479.

Darwin's attention. For Darwin, it was potentially an important meeting. He wanted to explore his evolutionary ideas with the botanists and zoologists whom he considered most knowledgeable about the variability of species.[22] Had Watson come, he could have joined Darwin's inner circle of trusted confidants. Instead, he became a member of Darwin's vast network of science correspondents who were happy to pause in their own researches long enough to consider questions relating to their expertise. Interacting by mail was better for maintaining their relationship, for Watson's sarcasm was probably less likely to appear in his letters to Darwin than in spontaneous conversations.

Darwin had already resolved his questions on the Azores by then, using Watson's articles and Hooker's advice. In the 15 letters that Watson wrote between 5 June 1856 and 23 March 1858, the discussions were mostly answering Darwin's questions about the species and genera of British plants, with some discussion also on the similarities between British and North American species. Darwin compared Watson and Asa Gray's judgments on the similarities between British and American species, and one of Watson's letters was written to Darwin in a format for forwarding to Gray.[23] Two recent studies explain how Darwin hoped to use the data of these letters to show that varieties are incipient species and to develop his principle of divergence.[24] However, Darwin drew upon the researches of more naturalists than Watson, Hooker, and Gray to reach his conclusions, and neither of these studies makes clear how much effort Watson expended in answering Darwin's questions. To appreciate that, one must read the actual letters in the magnificent compilation of Darwin's *Correspondence*.[25]

Watson seconded (5 June 1856) Lyell's advice that Darwin should proceed with publication, pointing out that improvements could be published later. In the same letter, he repeated his reassurance that 'You can make such use as you deem fit of answers to questions which you pay me the compliment of asking', but Watson was a little embarassed that his 'replies are unavoidably of that vague & unsatisfactory kind which must render them of no really available use'.[26] Darwin, of course, was in a better position to judge that matter than Watson. Despite Darwin's well-formulated questions and Watson's crisp, detailed answers, Watson persisted in thinking that the monograph which Darwin was busy writing could only be written in the distant

---

[22] The date and the persons invited are mentioned in a letter to T.H. Huxley, 2 April [1856], Darwin 1985–, VI, 66–7. On the importance of that meeting to Darwin, see Browne 1995, 538–40; Desmond & Moore 1991, 434–5.

[23] Watson to Darwin & Gray, 13 March 1857; Darwin 1985–, VI, 355–8.

[24] Browne 1980. Parshall 1982. See also Ospovat 1981, 179.

[25] There is also extant a draft of Darwin's letter to Watson, apparently sent in mid-June 1856; see Darwin 1985–, VI, 140–1.

[26] In Darwin 1985–, VI, 132.

future. This misperception persisted even after an exchange of letters on Darwin's article 'On the Action of Sea-Water on the Germination of Seeds'. Watson agreed that many botanists had been too pessimistic about the possibility of seeds remaining viable after floating great distances in the oceans, but wondered if Darwin did not go too far in the other direction: Darwin conducted his experiments in still water at shallow depths, which seemed to Watson to be overly favorable circumstances. Darwin countered with the hypothesis that many seeds which might not float freely for long periods might nevertheless be carried great distances caught among the roots of floating plants. Watson admitted that he had not thought about this possibility.[27] A few months later, he wrote to Combe (3 February 1857) stating that Darwin was writing a book that would address questions on successive development, and he expressed optimism about it:

> I anticipate that he will treat [species questions] in a juster manner than other scientific celebrities have done. I do not quite know what his work is. He has written to me several times about it, not in explanation, but merely with questions which he required to be answered by a botanist; his line, as you are likely aware, being Geology and Zoology.

Darwin was so enthusiastic about Watson's findings that he was eager to read volume IV of *Cybele Britannica*. Watson realized that Darwin's expectations were too high, and gave his modest reply that he was only hewing stones for a later edifice. Darwin was not put off by this attempt to lower his expectations, and Watson therefore sent along some extracts from that volume which related to Darwin's inquiries.[28]

Watson's Own Scientific Conclusions

Although the introduction to *Cybele Britannica*, in volume I, contained a brief version of some of Watson's phytogeographical ideas (and his notorious attack on Edward Forbes[29]), the most elaborate presentation ever of his scientific conclusions appeared in

---

[27] Watson to Darwin, 10 and 19 November 1856; Darwin 1985–, VI, 265–6 and 276. Darwin 1857 was read at The Linnean Society on 6 May 1856 and was obviously printed by November – before the entire volume was completed in 1857. For Darwin's investigations on this subject, see Browne 1983, chap. 8, and 1995, 517–21; Desmond & Moore 1991, 443–4.

[28] The list and explanations are quoted in Darwin 1985–, VII, 55–7, along with Watson's letter (23 March 1858) which offers further explanations; ibid., 53–4. Cf. Watson 1847–50, IV, 39–43. For Watson's letter lowering expectations, see above, n. 21.

[29] Watson 1847–59, I, 55 and 465–72. This relationship is discussed above, chap. 9.

volume IV. In the introductory chapter to that volume, he lamented the paucity of recent works on the distributions of particular plant species, acknowledging that the works of Humboldt, Meyen and Brown were useful in their day, but were now obsolete due to their vagueness. Watson probably saw the four annual reports on progress in botanical geography by August Heinrich Rudolph Grisebach which had been translated into English during the 1840s, but he need not have commented on such ephemera, which were only a faint foreshadowing of Grisebach's great work, *Die Vegetation der Erde nach ihrer klimatischen Anordnung* (1872).[30] Arthur Henfrey's modest survey, *The Vegetation of Europe, Its Conditions and Causes* (1852), devoted some ten pages to explaining the methodology and data in *Cybele Britannica*. Watson could therefore call it a 'pleasing volume', though grumbling that Henfrey had not always acknowledged his debts elsewhere in the book when using Watson's publications. (One wonders whether Watson bothered reading to the end, where he would have found a generous account of Forbes's paleo-phytogeographical studies.[31])

Watson mentioned a few other minor authors, but he only took seriously the phytogeographical works by Joseph Dalton Hooker and Alphonse de Candolle. Although he appreciated the magnitude of Hooker's achievements and his willingness to seek Watson's advice, since Hooker did not publish directly on the British vegetation, Watson mainly complained about Hooker's defenses of both Forbes and the stability of species.[32] Watson seems not to have known Henri Lecoq's *Études sur la geographie botanique de l'Europe* (9 vols, 1854–58), though Lecoq used both Watson's *Remarks* (1835) and *Cybele*.[33]

This left only Alphonse de Candolle (1806–93, son of Augustin-Pyramus de Candolle) to act as a foil for much of Watson's discussions, and it was to his monumental *Géographie botanique raisonnée* (1855) that Watson responded.[34] Their

---

[30] Humboldt and Brown are discussed above, chap. 2. Franz Julius Ferdinand Mayen's *Grundriss der Pflanzengeographie* (1836) had appeared in English translation in 1846. On Mayen, see Stafleu & Cowan 1976–92, III, 438–9. Grisebach 1846a and 1846b, 1849a and 1849b, 1872a. On Grisebach, see Stafleu & Cowan 1976–92, I, 1007–11; Wagenitz 1972.

[31] Watson had either forgotten Henfrey's earlier defense of the fixity of species or considered it inconsequential. Henfrey 1852, 161–70 on Watson; 378–84 on Forbes; Henfrey 1849 on fixity of species. On Henfrey, see Desmond 1977, 301; Geison 1972; Rehbock 1983, 148–9; Stafleu & Cowan 1976–92, II, 154–5.

[32] Hooker solicited Watson's help when comparing British species with those of the Southern Hemisphere in *Flora Antartica* (3 vols in 6, 1844–60). Watson 1847–59, IV, 9 on Hooker's defense of Forbes; IV, 33 and 50 on Hooker and species. On Hooker's publications, see Desmond 1977, 318; Stafleu & Cowan 1976–92, II, 267–83.

[33] I am endebted to Jean-Marc Drouin for pointing out Watson's neglect of Lecoq and for Lecoq's citations of Watson's works in Lecoq's II, 377, and III, 83. On Lecoq, see Stafleu & Cowan 1976–92, III, 804–5.

[34] On Alphonse de Candolle, see Browne 1983, 82–95; Christ 1893; Dajoz 1984, 13–20; Pilet 1971a; Stafleu & Cowan 1976–92, I, 433–7.

communications had begun on 23 February 1846 when de Candolle sent some botanical publications in hopeful exchange for a copy of Watson's *Remarks on the Geographical Distribution of British Plants* (1835). Watson gladly obliged him and wrote (14 August) of his plans to publish *Cybele*. An illustration of their exchange of views is Watson's comment (6 October 1848) on de Candolle's article, 'Mémoire sur les causes qui déterminent les limites des espèces du règne vegetal,'[35]: some 'mode of comparing climate & vegetable limits, as that you suggest, is needful, I am quite satisfied. Still, as you yourself see, there will be many exceptions and modifications required'. Watson's example was that although *Alyssum calycinum* [= *A. alyssoides* L., Hoary or Small Alison] was imported into Britain with seeds of *Trifolium pratense* [Red Clover] and *Medicago falcata* [Yellow Medick], it had propagated itself in Scotland, showing that Britain should be included within its range of climate. We may assume that Watson sent de Candolle copies of the *Cybele* volumes when they appeared, and that de Candolle was only reciprocating when he sent Watson the two volumes of *Géographie botanique raisonnée* which Watson acknowledged on 28 October 1855.

It was obvious to both de Candolle and Watson that any comprehensive discussion of the distribution of species should address the question of what constitutes a species. Although admitting in Chapter 2 some doubt about the reality of species, Watson quoted in his own English translation de Candolle's definition of a plant species.[36] Watson raised some objections, then offered his own definition:

A botanical *species* is a collective aggregate of individual plants, closely resembling each other (*firstly*) in having certain definable characters, common to all of them, by which any of them may be recognized, and by some of which they may be distinguished from any other plants; – (*secondly*) the same characters being repeated in their descendants, during successive generations, for a protracted and indefinite time; – (*thirdly*) the individuals not being always strictly alike; but such variations as do occur, being apparently inconsonant and reconvertible; that is, either known to be so from observation, or inferred to be so from analogy.[37]

He could imagine objections to his definition as readily as to de Candolle's, and he also discussed them.

Watson had already demonstrated in 1845 the difficulties in determining particular species by indicating how many species six British authors recognized

---

[35] Candolle 1847a.
[36] Watson 1847–59, IV, 29–30.
[37] Watson 1847–59, IV, 31–2.

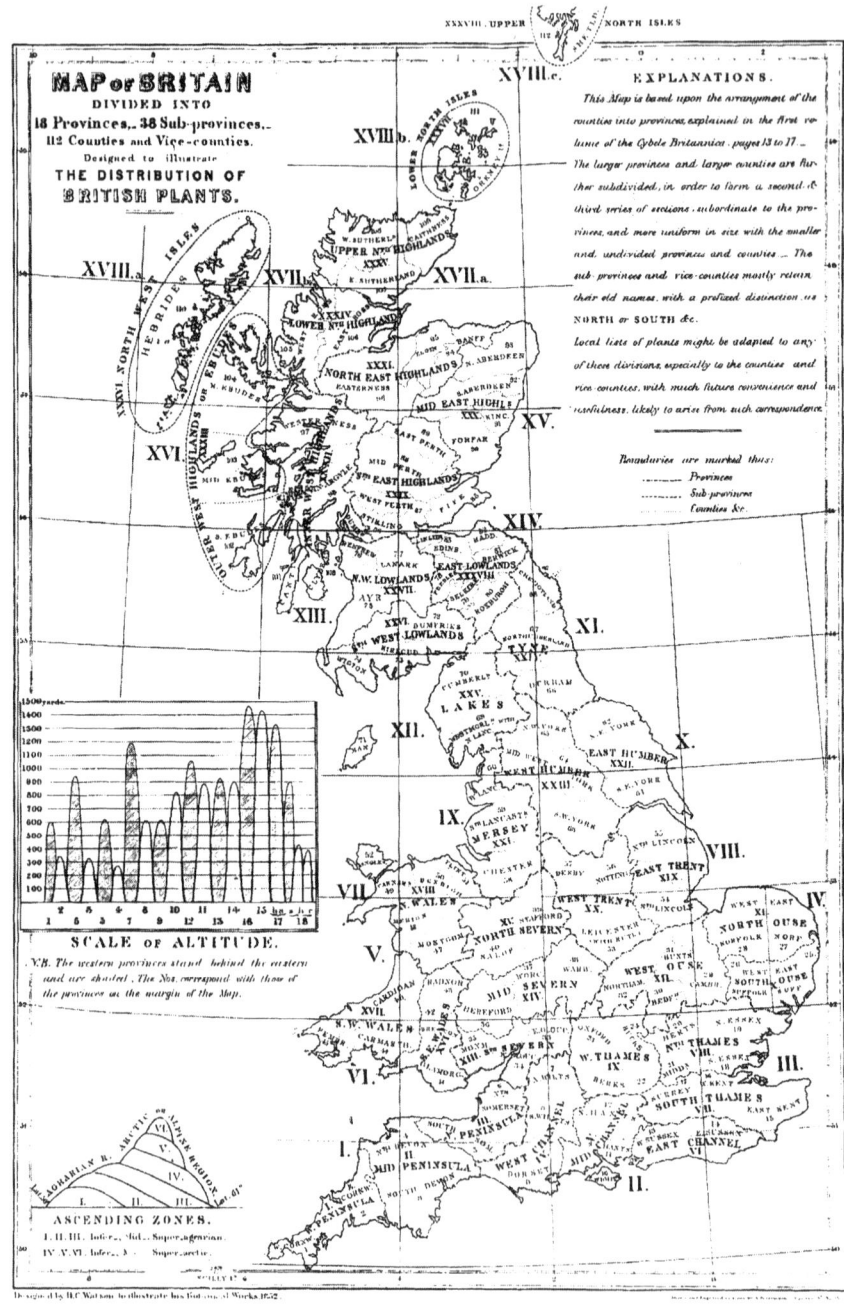

Figure 13 Provinces, sub-provinces, counties and vice-counties of Britain. Watson (1847–59).

within five genera, and in 1859 he expanded this discussion to include more authorities and more genera.[38] The increasingly detailed knowledge which botanists had acquired of British species during the fourteen intervening years had increased, not decreased, their disagreements. Some species appeared quite distinct and not subject to dispute, but others seemed indistinct, with various subspecies. Clearly, the distributions of species could not be ascertained adequately if their specific names meant one thing here and another there.[39] The same problem could be explored temporally as well as geographically by consulting the fossil record. Charles Lyell had done so in *Principles of Geology*, yet he seemed unwilling to apply the same reasoning to explain gradual change to the organic world that he did to the inorganic world. Watson suspected that a religious bias might have inhibited him.[40] Watson reaffirmed his own 1845 arguments in the *Phytologist* on the gradual change of species, citing the reference and quoting his earlier conclusions.[41] (His thoughts on the subject had not progressed much since then.) He ended this discussion by summarizing ten points he had just made concerning the gap between species in nature and in the scientific literature. Two examples illustrate his thinking:

3   Technical species, or those described in books by technical botanists, are theoretically supposed to correspond with the natural species.
4   But very few of such technical species are ever subjected to those tests which are deemed adequate to establish natural species.[42]

Watson's discussion of species in volume IV is one of the more sophisticated of the pre-Darwinian period. However, it shows the thinking of a skeptic rather than of a theoretician. He exposed the weaknesses of the opponents of evolution, and mentioned some of the evidences for the gradual change of species, but he was unable to develop a new theory and marshal the evidences for it in the way that Darwin would. For one thing, Watson was primarily committed to phytogeography (his word, in this volume). He lacked Darwin's determined, persistent focus upon 'the mystery of mysteries'. Even after he had read Darwin's and Wallace's 1858 announcements of their similar theories of evolution by natural selection, Watson was a little slow to understand how they had provided a new foundation for phytogeography. However, he was still ahead of everyone except Darwin and Wallace when he wrote to Alphonse de Candolle (8 June 1859): 'The more I think about the views of C. Darwin & Wallace,

---

[38] Watson 1845f, 143; 1847–59, IV, 39–44.
[39] Watson 1847–59, IV, 49–50.
[40] Watson 1847–59, IV, 53–5. Lyell 1830–33, II, 129, 181–2. Egerton 1968, 234–40.
[41] Watson 1847–59, IV, 59–63.
[42] Watson 1847–53, IV, 63–4.

the more disposed I feel towards belief that they are in the right course to explain the changes of species from eras past to eras present.' Competition played a key role in the distributional explanations of Augustin-Pyramus de Candolle, Lyell, and in Darwin's *Origin*,[43] but Watson overlooked its importance. This might have something to do with his following the lead of Alphonse de Candolle. The latter was familiar with his own father's writings, and one scholar argues that he also expressed some awareness of the significance of interspecific competition for plant geography,[44] but his emphasis was on the importance of inanimate environmental factors, and his discussion of competition was too inconspicuous to catch Watson's attention.

Watson's third chapter addressed the question of whether one could discover which species are native to Britain and which were introduced through human action. De Candolle had made a good start by using *Cybele Britannica*, vols I–III; unfortunately, he had also used other published records for Britain which Watson considered unreliable.[45] Watson easily demonstrated his superior knowledge, at least for Britain, with his technique of carefully scrutinizing large amounts of evidence. So far, he had discussed topics of general interest to botanists, but Chapter 4, which was a fifty-page discussion on the relationship between species distributions and geography and climate, was too detailed for most botanists to absorb unless provided with some reason to do so. Chapters 5–9 were even less accessible, for there was less narration, and there were more lists of species compiled from various perspectives. His methodology exposed him to two risks: first, that he would merely redo De Candolle's work for Britain, achieving only a higher degree of precision, and second, that he would merely compile lots of data from various viewpoints which might be useful, but within a context which he had not developed.

Chapter 10, 'General Remarks', contains useful conclusions, yet the one that most intrigues modern ecologists did not elicit much response at the time. In 1835 he had astutely estimated: 'On the average, a single county appears to contain nearly one half the whole number of species found in Britain; and it would, perhaps, not be a very erroneous guess to say that a single mile may contain half the species of a county.' By 1859 he had more data and could show that a square mile in Surrey had nearly half the species of the county (400 out of 840 species). Carrington B. Williams commented on this finding in 1943, and it has continued to attract notice.[46]

---

[43] Egerton 1968, 1971, 1974.
[44] Dajoz 1984, 41.
[45] Watson 1847–59, IV, 83–107. Watson's arguments here are too complex for easy summarization. The work which he felt was least reliable for de Candolle's purposes was the posthumous *Flora Vectesis* (1856) by William Arnold Bromfield (1801–51). On Bromfield, see Desmond 1977, 91.
[46] Dony 1963, 377 and 383–4; Watson 1835a, 41–2; Watson 1847–59, IV, 376–83; Williams 1943 and 1964. See also chap. 6, n. 9.

Volume IV contains neither a new paradigm for phytogeography nor a tightly reasoned defense of a theory. Watson himself was aware of his not being able to provide a path-breaking treatise, and warned Darwin not to expect too much:

> You seem to expect something from the concluding volume of Cybele, – & will thus be disappointed. It will simply be like a block of stone, cut & fashioned in a particular manner, but by itself of little or no use; – yet adapted to constitute part of an edifice where other blocks are prepared in like manner, to be built up along with it. I can in some measure see or guess what ought to be done, – but am unable to do it, because the like materials do not elsewhere exist, for comparison & generalization.[47]

Premature conclusions are seldom useful, but this confession of incompleteness is an indictment of Humboldtian science, which was weak in theoretical methodology.

Although Watson sent gratis copies of his books to important botanists, he declined to send copies to editors of periodicals, because 'no desire was felt for seeing the book reviewed by persons who had given much less attention to its subject than [had] the author ...'[48] Consequently, only two reviews of the entire *Cybele Britannica* are known: one well informed and mostly positive, the other ill informed and negative.

Fortunately for Watson's self-esteem, Alphonse de Candolle's favorable one, in French, appeared first, in July. He noted that *Cybele Britannica* was the first published botanical geography of a country, that its level of accuracy seemed high, and that 'It is desirable to have works of this character for other countries, as complements of their Floras, and as means of comparison in botanical geography.'[49] He admitted that, as a foreigner, he may have accepted published British reports on species distributions that were not credit worthy, but in turn he criticized Watson for demanding precision where it could not readily be achieved and for avoiding use of hypotheses where they could advance science. Watson graciously acknowledged (4 September 1859) receiving de Candolle's review with the comments that 'Our views and ideas differ on some points. But such difference does not interfere with the very real respect with which I look up to you, as very far my superior in botanical knowledge.' Some months later Watson published his own translation of de Candolle's review, in which he responded to its criticisms with a rather pompous phrenological discourse on the mental attributes of botanists – a rehash of his 1833 remarks.[50]

---

[47] Watson to Darwin, 3 January 1858; Darwin 1985–, VII, 1.

[48] Watson 1860, 6.

[49] De Candolle 1859, 273 (I am indebted to Jean-Marc Drouin for a copy of the original review); translated in Watson 1860, 11.

[50] De Candolle's criticisms are in Watson 1860, 12–13. Watson's response is on 18–25. Cf. Watson 1833c.

In November a hostile, anonymous review appeared in *The Gardeners' Chronicle*. John Lindley, Professor of Botany at University College London, was horticultural editor of that periodical, and author of the review. He acknowledged Watson's enormous amount of labor, but dismissed the results as inconsequential: 'Instead of precise results we have elaborately learned disquisitions, which really, when dissected, end in nothing.'[51] This was partly sincere conviction that Watson had failed to advance botanical geography beyond what had been achieved by Brown, Humboldt and Alphonse de Candolle, but partly also a slap at Watson's rudeness in print toward others, and maybe partly a retaliation for Watson's mixed review of Lindley's *Elements of Botany*.[52] Watson responded with sarcasm the following year in *Part First of a Supplement to the Cybele Britannica* (p. 24).

Darwin's Use of Watson's Findings

British botanists had learned before 1859 that they ignored Watson's publications at their peril; despite Lindley's review, they would have scanned *Cybele* for whatever seemed relevant to their work. Nevertheless, the British scientist who most appreciated *Cybele Britannica* was not even a botanist. The ways in which Darwin planned to use Watson's information and conclusions are seen in the large manuscript entitled Natural Selection which he began writing on 14 May 1856 (not waiting for Watson's encouragement to do so, which was received three weeks later). In the first place, he adopted Watson's strategy of dealing with the species concept, which 'allows a naturalist to communicate a theory of the evolution of species to his fellow naturalists, even when the latter employ a non-evolutionary definition of the term "species"'.[53] More particularly, he discussed and cited Watson's botanical evidences and judgments on 27 different pages, especially in his chapter on variation in nature.[54] Watson would have been very gratified had Darwin published this work, which, in the event, did not appear until 1975.

Darwin's publishing plans changed after Alfred Russel Wallace sent him his own paper on natural selection in 1858. Their joint writings on natural selection were then read before the Linnean Society of London and Darwin began abridging

---

[51] [Lindley] 1859. The reviewer's identity was revealed by J. Hooker to Darwin on 21 November 1859, quoted below. On Lindley (1799–1865) see Desmond 1977, 386, Stafleu & Cowan 1976–92, III, 49–60 and Stearn 1973. A hostile review by zoologist John Fleming of Watson's volumes I–III (1854) shows Fleming was not a competent botanist and was a friend of Edward Forbes.
[52] Watson 1847e; but see also Watson 1847–59, IV, 16, 20–1.
[53] Beatty 1985. Hodge 1977 and 1987.
[54] Darwin 1975, see listing in index, 689.

Natural Selection into a much less heavily documented form: *On the Origin of Species* (1859). Since Darwin had consulted Watson about writing *Natural Selection* in 1856, he most likely also informed Watson about the desirability of abridging it into a more quickly publishable format (and if he did not, Watson would have read the story in *The Origin's* introduction). Watson had given Darwin a copy of *Cybele Britannica* when volume IV was published, and Darwin later reciprocated by sending him a copy of *The Origin*. Even in this compressed format, Darwin still managed to mention Watson's work on eight pages and acknowledged 'Mr. H.C. Watson, to whom I lie under deep obligation for assistance of all kinds ...'[55]

This generous expression enabled Watson to read *The Origin* with positive feelings, and he found it was an exhilarating experience. As soon as he finished, he shared his feelings about it in an enthusiastic letter to Darwin (21 November 1859):

Once commenced to read the 'Origin' I could not rest till I had galloped through the whole. I shall now begin to re-read it more deliberately. Meantime I am tempted to write you the first impressions, not doubting that they will in the main be the permanent impressions.

1st. Your leading idea will assuredly become recognized as an established truth in science, i.e. 'natural selection'. (It has the characteristics of all great natural truths, clarifying what was obscure, simplifying what was intricate, adding greatly to previous knowledge.) You are the greatest Revolutionist in natural history of this century, if not of all centuries.

2nd. you will perhaps need in some degree to limit or modify, possibly in some degree all to extend your present applications of the principle of 'natural selection' ... it strikes me that there is one considerable primary inconsistency, by one failure in the analogy between varieties & species; another by a sort of barrier assumed for nature on insufficient grounds, and arising from 'divergence'. These may, however, be faults in my own mind ...

Watson added a poignant historical note:

A quarter century ago you & I must have been in something like the same state of mind, on the main question. But you were able to see & work out the quo modo of the succession, the all-important thing, while I failed to grasp it. I send by this post a little controversial pamphlet of old date, – [Watson's polemic of 1836 on] Combe & Scott. If you will take the trouble to glance at the passages scored on

---

[55] Darwin 1859, quotation 48; the other mentions are on 53, 58, 140, 176, 363, 367 and 376. Only six of these references are indexed in the 1859 edition, and none are included in the revised index for the 1964 facsimile edition.

the margin, you will see that, a quarter century ago, I was also one of the few who then doubted the absolute distinctness of species & special creations of them. Yet I, like the rest, failed to detect the quo modo which was reserved for your penetration to *discover*, & by your discernment to *apply*.

This letter ends with speculations which Darwin's writings had stimulated concerning human evolution: Watson suggested that a superior species 'would make direct & exterminating war upon his Infra-home cousins'.

The very next day Darwin received both Watson's letter and one from Hooker answering a question which Darwin had posed on 20 November: who wrote the severe review of *Cybele Britannica*, volume IV, in the *Gardeners' Chronicle*? Hooker, who occasionally recommended Watson's work to Darwin, responded on 21 November 1859:

Of course Lindley wrote the G.C. article on H.C.W. & upon my honor I do think it richly deserved. The sneering contempt with which [Watson] treats his enemies, the virulence of his dishonest attacks on those he knows little of, & his patronizing air to those he approves, are beyond all whipping powers of reviewers ...

In response, then, to both Watson's and Hooker's letters of the same date, Darwin wrote to Hooker (22 November):

I cannot help being sorry about H.C. Watson: he has helped me so kindly & liberally. I had a letter from him with such tremendous praise of my book, that modesty (as I am trying to cultivate that difficult herb) prevents me sending it [to] you, which I sh^d have liked to have done, as he is very modest about himself.[56]

Reading *The Origin* helped Watson ignore the hurtful review in the *Gardeners' Chronicle*, though he thanked Darwin for his comments on it, claiming that improving Darwin's book was more important than reading reviews of his own (30 November):

Page 49 –
    Instead of 'Primula veris & elatior'
    Print 'Primula vulgaris & veris'.
    Explanations wherefore, if wished, are over leaf
        vulgaris is the name of *Primrose*

---

[56] These letters are in Darwin 1985–, VII: Watson to Darwin, 21 November, 385–6; Darwin to Hooker, 20 November, 382; Hooker to Darwin, 21 November, 383 with notes on 384; Darwin to Hooker, 22 November, 387 with notes on 388.

veris is the name of *Cowslip*

elatior is the name of **two different** things, neither Primrose nor Cowslip: –

1ˢᵗ (& correctly) the name of a species, different from P. & C., without any safe evidence to show that it passes into either.

2ᵈ. (by misnomer originally) the name of an intermediate variety between Cowslip & Primrose, & producing both these from its seeds.

The earlier recorded experiments are of doubtful reliance, because their recorders either do not explain, or did not know, to which elatior their seeds or individuals belonged.[57]

Since Darwin was not a botanist, Watson neither chastised him for overlooking what Watson had already published on this subject nor even suggested that Darwin cite his own discussions of it.[58]

---

[57] Darwin 1985–, VII, 408. Cf. Darwin 1959c, 132.
[58] Watson 1841a, 1845h, 1845i, 1847c, 1848n and 1847–59, II, 292–3.

PART III
LATER LIFE, WORK AND INFLUENCES

Chapter 14

Later Life, Work and Influences, 1860–81

> It was my wish and expectation through years, to write a work in criticizing review of the various hypotheses suggested to account for actual distribution ...
> Watson to de Candolle, 31 January 1874

> ... the journal's title ... was to be *Watsonia*, in commemoration of the man whose particular set of interests the Society faithfully embodies and to whom it substantially owed its very survival.
> David E. Allen 1986, 140–1

Watson had worked toward a revolution in biology since 1834, and he was the first to say that Darwin had achieved it. He was thrilled to see it arrive, and one might imagine, therefore, that it would be a turning point in his life. But Watson was an observer and a traveler down his own path, not someone else's. In 1860 Watson turned 56 – not old by modern standards, but in the Victorian era, old enough to begin putting one's affairs in order in preparation for old age. Indeed, he seems to have believed he was already old, for in declining an invitation to visit the Trevelyans, he explained that: 'throughout life I have been a shy recluse, & now health has become so uncertain as to render me still more so' (1 August 1860). Although Watson never became a university professor, he did attract a protégé to whom he could pass on some of his knowledge and wealth. He reflected upon his experiences, gained in self-knowledge, and continued his productive life until a few months before death.

Doubts on the Darwinian Revolution

If Watson was Darwin's quickest convert, he was not his most loyal disciple. In 1860 Watson was willing to claim in print that the *Origin* was 'seemingly the most important volume on natural history ever published',[1] but his equivocal 'seemingly' continued the provisional reservations raised in his first letter of enthusiastic

---

[1] Watson 1860, 30.

commentary (21 November 1859). That letter had soon been followed by another with a minor correction.[2] These were appropriate responses, which Darwin had welcomed. Darwin produced in the *Origin* a new paradigm, a reinterpretation of a science which also sets the agenda for new researches.[3] Watson could have spent the rest of his research career testing, expanding and revising the Darwinian paradigm. However, he was five years older than Darwin, and his life's work already had momentum; both Hooker and Huxley, whose subsequent work was heavily influenced by the *Origin*, were several years younger than Darwin. It would be fairer to compare Watson not to these disciples, but to Richard Owen, who was Watson's age and who never accepted Darwin's version of evolution, preferring to believe that if evolution occurs, then he had figured out how it happens himself. Owen's judgment seems distorted by envy.[4]

In 1860 Watson continued working with Darwin on potential revisons or refinements for later editions of the *Origin*.

About 3 January 1860 Watson had reread the first edition and sent Darwin a 14-point argument on a hypothetical difficulty with his theory – that continual divergence of variant offspring would lead to ever-increasing numbers of species beyond what really exist. Watson thought 'that the difficulty could be met and neutralized by the hypothesis or inference that individuals *converge* into orders, genera, species, *as well as diverge* into species, genera, orders, through nepotal descent'.[5] He understood that Darwin had rejected this idea in the *Origin*, and that it would be difficult to find supporting evidence. Nevertheless, Darwin found Watson's problem interesting, admitted that he also had been bothered by it, and gave a detailed answer including evidence not presented in the *Origin*.[6] Darwin seems to have sent Watson a copy of the revised third edition (1861).[7]

Before the third edition was off the presses in April, however, Watson sent Darwin a letter which was not saved. Darwin reacted to it in a letter to Hooker:

---

[2] Watson's endorsement (1847–59, IV, 388) of Darwin's 'remarkable and profound views', and his letter to Alphonse de Candolle (8 June 1859) are evidences that he may have been Darwin's quickest convert. His letters to Darwin in November 1859 are quoted above, pages 191–3.

[3] On the application of Kuhn's ideas to the Darwinian revolution, see Greene 1981, chap. 3, Mayr 1972b, and Reingold 1980.

[4] Their respective life dates were: J. Hooker (1817–1911), T.H. Huxley (1825–95) and Owen (1804–92). On Huxley, see Williams 1972. On Owen, see Desmond & Moore 1991 passim (see index), Rupke 1994, Ruse 1979 passim (see index) and Williams 1974.

[5] Darwin 1985–, VIII, 10–13, quotation on 12. Cf. Watson 1860, 26–30, on the status of genus in botany.

[6] Darwin 1985–, VIII: Watson to Darwin, [3? January 1860], 10–13; Darwin to Watson, a partial draft of letter [5–11 January 1860], 17–19.

[7] Darwin 1985–; Watson to Darwin, 10 May 1860, VIII, 202–4; list of recipients of third edition, IX, 420–1 and 424.

Here is a good joke: H.[C]. Watson (who I fancy & hope is going to review new Edit. of *Origin*) says that in first 4 paragraph[s] of the Introduction, the words 'I' 'me' 'my' occur 43 times! I was dimly conscious of this accursed fact. – He says it can be explained phrenologically which I suppose civilly means that I am the most egotistically self- sufficient man alive ...[8]

Darwin had sent Watson four published reviews of the *Origin* to read, and when Watson returned them he probably mentioned (in a lost letter) writing his own review of the *Origin*, and if so, Darwin naturally assumed he would publish it soon and in a periodical – both assumptions were wrong. Although Darwin's species theory was more sophisticated than Watson ever realized, Darwin never published the manuscript which directly explained his species theory in greatest detail.[9]

Darwin had a persistent interest in the judgments of botanists on particular species and varieties and their geographical distribution. Watson was such an authority for British vascular plants (second only to Babington), and Darwin sent him further queries on 17 July 1861. In his response, Watson called attention to a discussion on 'the dubiety of the records, arising from the diverse applications of the same name' in the first *Supplement to the Cybele Britannica*, a copy of which he had sent earlier.[10] The queries, however, continued in both directions. In the same letter, Watson wondered: 'Why cannot two varieties be fertilized differently, as well as two species?' Only half convinced, he found a partial answer in the *Origin*: '... you are probably right; because by species you really mean only wider varieties or races, and each additional difference helps to widen'. A day or so later, Watson sent Darwin a postscript which shows further indecisiveness where both men would have preferred decisiveness:

P.S. As to Veronica humifusa & Serpyllifolia. – Some botanists hold the former a good species. And the facts read either way.

V. humifusa perishes under given cir[cumstan]ces where V. Serpyllifolia flourishes, & naturally occurs under diff· cir[cumstan]ces.

Therefore it is a species of itself. – Or, therefore a var[iet]y has a different climatal adaptation – Or, therefore a var[iet]y is caused by difference of climate & soil acting on a succession of generations.[11]

---

[8] Darwin to Hooker, 27 [March 1861], Darwin 1985–, IX, 70–1.
[9] Darwin 1975; Hodge 1977 and 1987.
[10] Darwin to Watson, [17 July 1861], Darwin 1985–, IX, 207–8; Watson to Darwin, 24 July 1861, IX, 217–18. Watson's reference to Darwin was 'to the middle paragh. of page 45 & examples in sequence', in Watson 1860.
[11] Watson to Darwin, [after 24 July 1861], Darwin 1985–, IX, 219.

Watson's methodologies were oriented toward detecting and describing anomalies, but were inadequate for resolving them. Darwin aspired to resolve them to provide evidence for his theory, and his methodologies were more wide-ranging. Watson did the fieldwork and compared specimens to each other and to published descriptions, whereas Darwin attempted to explain the anomalies after they were discovered. In other words, Darwin began where Watson left off. A good example was the anomaly of British primroses. Watson had attempted to sort out the different forms in the 1840s, and Darwin had mentioned the anomaly in the *Origin* (only to have Watson send him a correction after its publication). In 1860 Darwin began reading the literature, writing to botanists, and conducting his own experiments on four species of Primula: *vulgaris, veris, sinensis* and *auricula.* Darwin felt uncertain about the reliability of experiments which Joseph Sidebotham had published in 1849, and Watson's detailed letter of 20 September 1861 both explained why he distrusted Sidebotham's report and gave his own views on the questions raised.[12] Darwin's own paper on *Primula* species was nearly finished, and the only information from Watson's letter which he seems to have needed was confirmation of his suspicion about Sidebotham's experiments. This might explain Darwin's failure to acknowledge Watson's letter in this article. On the other hand, it might just have been an accidental oversight.[13]

The third edition of the *Origin* (1861), which Darwin sent to Watson, contained a historical introduction which also failed to mention Watson – despite Watson having called Darwin's attention to his own publications on evolution beginning in 1836. The omission may have been just an accidental oversight, but one might wonder whether Watson's gratuitous observations about Darwin's overuse of 'I' 'me' and 'my' might have caused Darwin unconsciously to 'forget' this critic when acknowledging other precursors. It is noteworthy that Watson, uncharacteristically, did not complain about these slights in any of his later writings. This was fortunate because Darwin, in three of his later books – *The Variation of Animals and Plants Under Domestication* (1868), *The Effects of Cross and Self Fertilization in the Vegetable Kingdom* (1876), and *The Different Forms of Flowers on Plants of the Same Species* (1877) – did carefully acknowledge his use of Watson's writings. Furthermore, Darwin occasionally sought his advice and gave him copies of his publications.[14]

---

[12] Watson to Darwin, 20 September 1861, Darwin 1985–, IX, 270–2. Sidebotham 1849.

[13] Darwin 1862 (read 21 November 1861). Allan 1977, 265–9, seems unaware of Watson's then unpublished letter to Darwin on 20 September 1861 in her characterization of the available knowledge before Darwin published.

[14] In July or August 1862, Darwin requested his help in obtaining specimens of *Lythrum hyssopifolia*, and in the same year sent him a copy of his *Primula* paper and his book on orchids. Darwin 1997, 360–1, 669 and 677.

Nevertheless, Watson and Darwin were not regular correspondents during the next two decades, though they pursued researches that sometimes could have been interrelated. Watson continued to doubt the adequacy of Darwin's theory. Before their correspondence lapsed, Watson solicited the assistance of Rev. George Gordon in obtaining specimens of *Goodyera* for Darwin while expressing his own reservations about Darwin's theory (19 September 1860, 27 June 1861). And in a letter to the paleobotanist William Carruthers, Watson wrote:

I should very much like to see a Fossil Flora so treated as to show how far the full array of its facts really bears in favor of Darwin's theory. We can scarcely refuse the inference, that present species have in some way emanated from past species, but the modus inchoandi is the puzzle to me still. Darwin's 'selection' rather begs the question than truly answers it, – fits an explanation to the facts, and then declares that his explanation accounts for the facts. You have a knotty subject to unravel, if you pass onward from the descriptive to the explanatory.[15]

An amateur botanist to whom Watson confided his doubts happened also to be a leading philosopher. One of several areas of philosophy that interested John Stuart Mill was the philosophy of science. Although their link was botany, when Mill responded to Watson's doubts, he wrote in his capacity as a philosopher of science:

In regard to the Darwinian hypothesis, I occupy nearly the same position as you do. Darwin has found (to speak Newtonially) a *vera causa*, and has shewn that it is capable of accounting for vastly more than had been supposed: beyond that, it is but the indication of what may have been, though it is not proved to be, the origin of the organic world we now see. I do not think it an objection that it does not, even hypothetically, resolve the question of the first origin of life: any more than it is a by frequent blending. The difficulty is also met by the fact that the law of natural selection must cause all forms to perish except those which are superior to others in power of keeping themselves alive in some circumstances actually realized on the earth.[16]

---

[15] Watson to Carruthers, 18 June [?]. On William Carruthers (1830–1922), see: Desmond 1977, 120, and Stafleu & Cowan 1976–92, I, 460–1.

[16] Mill to Watson, 30 January 1869, Mill 1963–91, XVII, 1553–4, in response to Watson to Mill, 14 January 1869 (unpublished). Watson replied to Mill on 6 February and Mill responded to it on 24 February; Mill's second letter and parts of Watson's second letter are published in Mill 1963–91, XVII, 1567. *Vera causa* was a technical concept in philosophy of science; Ruse 1979, 56–63 et passim; Hull 2000. On Mill as botanist, see above, page 142.

Figure 14 H.C. Watson (1871). Photo by Maull and Fox.

Despite Mill finding some common ground between them, he was impressed mainly by the explanatory power of Darwin's theory, whereas Watson focussed mainly on alleged weaknesses. Watson and Mill exchanged these comments more than a year before Watson published his evaluation of Darwin's theory; however, their exchange was not in preparation for writing it. The reverse actually seems to have occurred: Watson had his commentary on Darwin's theory printed (but not published) in 1868, and it might have been while proofreading it that he decided to ask Mill for his assessment. Darwin often modified his writings in response to comments received from naturalists whose judgment he respected. Watson did so less often, and he found nothing in Mill's comments to cause him to modify what he had written.

Watson buried his evaluation of Darwin's theory under a subheading within the introduction to his *Compendium of the Cybele Britannica* (1870). Its near invisibility carries an implicit message that however revolutionary Darwin's theory might be, it would not have much impact on Watson's work. After all, he had been an evolutionist and phytogeographer before Darwin, and he may have believed that Darwin's ideas had only slight implications for Watson's project, begun four decades earlier. Since Darwin had not mentioned Watson in 'An Historical Sketch of the Progress of Opinion on the *Origin* of Species, Previously to the Publication of the First Edition of This Work' (1861), Watson remedied that oversight by quoting the transmutationist passages from *An Examination of Mr Scott's Attack upon Mr Combe's 'Constitution of Man'* (1836) and by reminding readers that he had already quoted from his 1845 transmutationist essays in *Cybele*, volume IV.[17]

Watson feared that the substantial support for Darwin's theory would lead scientists to assume that he had explained more than he had. Four weaknesses bothered Watson:

1 natural selection has two defects
   a it does not explain the causes of variation
   b *selection* is a human act of will and effort, and adding the prefix *natural* fails to change this.
2 Darwinian theory is based on the hypothetical assumption that variations can and do accumulate sufficiently to convert species into new species by divergences of characters gradually accumulated. But is there any change now noted that is sufficient to warrant the belief that a fern, a fir-tree, a moss and a mushroom ever had a common ancestor?
3 How had the original species arisen? Darwin stresses the extreme 'imperfection of the geologic record'.

---

[17] Watson 1870a: his discussion of Darwin's theory is on 43–59, which includes quotations from Watson 1836b on 46–8; his reference back to Cybele, vol. 4, is on 49. Darwin's historical sketch was added to the third and later editions of the *Origin*.

4   It seems a great or insurmountable deficiency that his theory 'is wholly one of constantly successive divergence from antecedent forms, without taking into account that divergence from one form, as now witnessed, usually is and must be approximate convergence towards ... some other form'.[18]

These were legitimate questions, and the challenge of any theory is to answer the doubts of skeptical scientists. Over the following decades, the Darwinians were indeed to provide answers, but meanwhile, during their lifetimes, Watson had various reasons for not giving Darwin wholehearted support.

A Synthesis on the Botany of the Azores

Watson, on his own, would not have resumed work on the plants of the Azores. He had little incentive to return there to collect specimens on the islands still botanically unexplored, for British botany could easily absorb all his scientific endeavors. Yet when newly collected specimens became available to him, he happily produced a critical synthesis on the botany of the Azores.

Frederick Du Cane Godman (1834–1919), an English naturalist educated at Cambridge, had a strong interest in exploring foreign lands. In the spring of 1865 he, a brother and an entomologist, Brewer, collected plants and animals on all the Azores Islands except Santa Maria, where Brewer collected alone. Afterwards Godman solicited the help of several specialists outside his own areas of expertise (mainly mammals and birds) to describe his specimens in a collaborative volume, *Natural History of the Azores, or Western Islands* (1870). In the 'Preface', he thanked all but one of his other collaborators in a rather perfunctory way, as if to say that their tasks were fairly modest in comparison to those of himself and one other; Watson's 175 pages were actually more than Godman himself wrote, though Godman's contributions were important for giving the book coherence. Godman commented that:

> It has been my especial good fortune to enlist the interest of Mr. Watson in my undertaking. The Botany of the islands, his study of which began during a personal visit, has continued for many years to engage his attention; and thus he was better qualified than any other botanist to deal with this portion of the subject.[19]

---

[18] Watson 1870a, 52–6.
[19] Godman 1870, iv. Despite his gratitude at the time, when he later supplied information for a biographical sketch on himself, the botanist's name was mistakenly published as 'Mr. Wilson'; anon. 1908, 86.

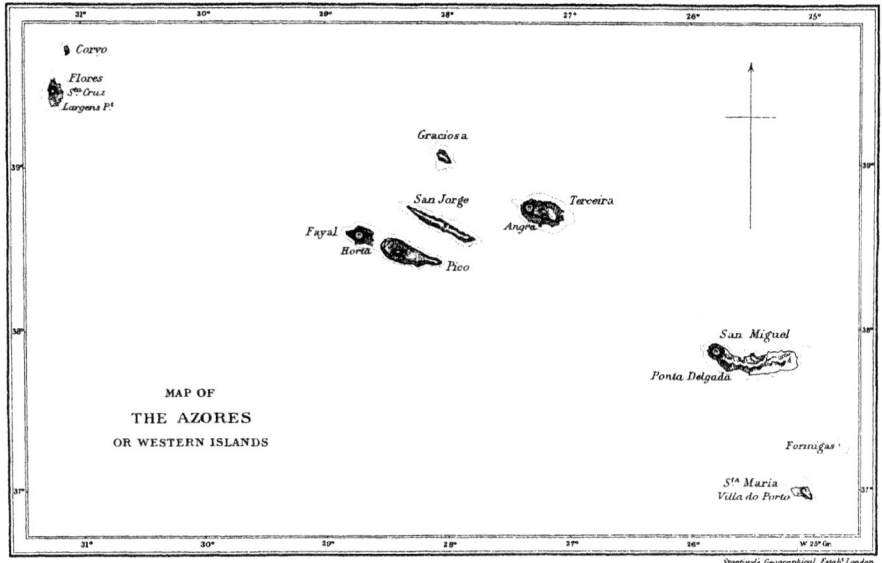

Figure 15 Map of the Azores Islands, Godman (1870).

Watson's decision around 1848 not to attempt a definitive work on Azores botany was now vindicated. New data made such an attempt more practicable than it had been earlier. His critical accounts of 478 species indicate their occurrence among the Azores and other Atlantic Islands. He then summarized his sources:

> Five collections have been combined to form the foregoing Catalogue; those of Hochstetter (*Flora Azorica*), Watson, Hunt, Godman, Drouet with his co-travellers of 1858 ... those of the several collections count up to the following numbers:
> Hochstetter, in Seubert's *Flora*   289
> Watson, 338; adding thereto   103
> Hunt, resident, 375; adding   67
> Drouet, Morelet, Hartung, 322; adding 13
> Godman, 256 or more; adding   6
>   478 species.[20]

---

[20] Watson 1870b, 262. Henri Drouet (b. 1829) was a French botanist and conchologist; Stafleu & Cowan 1976–92, I, 681.

At Godman's request, Watson listed all species with indications of whether each was known from Europe, Madeira, the Canaries, America and Africa. However, he cautioned that 'I have neither leisure nor inclination to fill in an American column properly.'[21] His list indicated that 40 species were known only from the Azores.

Watson saw the relevance of his new data for 'the Forbesian hypothesis of a great Continental extension westward or south-westward from Europe in its present limits, and he asked the rhetorical question of how well his data might support 'that hitherto utterly unsubstantiated conjecture?' His answer was that the Azores flora:

is European in its general character, and in detail it is numerously composed of European species; while nearly all of the additional species are found in the neighbouring island-groups of Madeira and Canary, or else are known only in the Azore Isles themselves. At first view, these facts may seem well in accordance with the hypothesis alluded to. And yet, in examining the more special details, they seem very difficult to reconcile with the idea of these Isles truly being the dissevered remnants of a great continental land formerly uniting them with Europe.

Furthermore, there was compelling counter-evidence:

The plants which must be held specially to characterize the Azore flora, at the present time, are precisely those which seem least fitted to endure a continental climate; being unable to bear any extremes of heat and cold, and especially dryness of climate.[22]

Since 1845, Watson had sought further evidence to discredit Forbes's theory, and now he displayed it in one of his best pieces of scientific reasoning. His excellent reasoning is an example of Karl Popper's conjecture-and-refutation prescription for doing science.[23] By 1870, biogeography was still mainly a descriptive science; Watson would have done better with more such provocative hypotheses to challenge.

He also used his data to test Darwinian ideas; he found both pro and con evidence. Two pairs of species provided positive evidence for Darwin's theory:

The shortened leaves and flowers, and the compact growth of the *Erica azorica*, might be held variations from the European *Erica scoparia*, in the direction which would better adapt the former to bear tempestuous winds, sweeping over the hilly surfaces of small islands; and we know that the climate of the Isles is

---

[21] Watson 1870b: the list is on 278–87; quotation 287.
[22] Watson 1870b, 273–5.
[23] Popper 1962 and 1965.

changeable and tempestuous. The procumbent habit of *Lysimachia azorica*, in contrast against the more prostrate and rooting habit of its near European ally *L. nemorum*, may also better adapt the former to its situations of growth, on the sloping banks and steep rocks of the Isles; nor is it any serious claim on credulity, to suppose these two to be really divergent descendants from a common ancestor, more or less intermediate between them.

But two other species did not seem to fit so neatly into Darwin's theory:

> The equability of temperature and the mild dampness of the Azore Isles are also conditions quite in natural correspondence with the evergreen and glabrous foliage of the *Veronica Dabneyi*, while the spreading wiry stems and the coriaceous texture of its leaves might also be held adaptations to its places of growth, on rocks much exposed to violent winds and subject to frequent showers. It is difficult in this case, however, to suggest any European *affine* or *analogue*, which could have diverged from the same single parental species; the divergence could only have been remotely ancestral. Still more difficult or impossible would it be, to name the European *Campanula* which could be accepted as a brother-species or cousin-species with the *Campanula Vidalii* of the Isles; although it is easy enough to look on this latter as a plant specially modified and adapted to its place of growth on coast rocks, under a mild and even temperature, but much subject to winter storms.

Darwin could only be pleased that these were the worst counter-evidences Watson's researches turned up. Darwin could easily retort that either the European analogues had never existed, or they had become extinct. Even Watson, when he balanced his positive against his negative evidence, saw that the positive was heavier.[24] This discussion appeared in the same year as Watson's more theoretical critique of Darwin's theory in *A Compendium*, and shows that Watson was better at reasoning from definite evidence than he was at theorizing.

A fundamental question interested Watson: what relationship exists between geographic distribution and variations in the forms of plants? He never developed adequate methodologies for answering it, however, and modern studies such as Briggs' and Walters' *Plant Variation and Evolution* (1969) have stronger roots in Darwin's studies made during the 1860s than in Watson's made over several decades. Still, Watson's studies probably helped Darwin see how to design his own researches. It has never become practicable to solve all such problems by experimentation. For example, on the basis of a few Azorean dried specimens, Watson in 1844 tentatively

---

[24] Watson 1870b, 276–7.

judged *Euphrasia grandiflora* Hochst. and *E. azorica* H.C.Wats. to be distinct species. However, after George Bentham reduced *E. azorica* to the synonymy of *E. grandiflora* in 1846, Watson accepted this judgment when publishing his final survey on the Azores flora in 1870. A century later, in 1973, P.F. Yeo, on the basis of still more herbarium specimens than had been available to either Watson or Bentham, restored *E. azorica* to the rank of species.[25] Watson and Darwin's interest in plant species of the Azores and their zonal distribution on Mount Pico has also been extended by twentieth-century phytogeographers.[26]

Relationship with Joseph Hooker

Watson's longest-lasting professional relationship was with Joseph Hooker. Since Watson's relationships with William Hooker, Babington and Balfour all deteriorated over time, why did this not happen also with Joseph? There were several reasons: on most botanical matters they had a high regard for each other's judgment; they shared an interest in Darwin's work, and as a public servant, Joseph felt obligated to maintain relationships that seemed of value for the public good.

In 1860, Watson cautioned Joseph that he tended to go too far with 'the current practice of infinite divisibility in species'(11 July), while Bentham went to the opposite extreme of 'lumping' distinct species (26 July). Concern over such determinations preceded publication of *Origin of Species* (1859), but Darwin's book placed the discussion within a more rigorous theoretical context than had previously existed. Watson shared with Hooker his notion that the divisions caused by natural selection into 'millions on millions' of species must be balanced by another tendency toward regression to earlier forms. Watson was pleased to tell Hooker that Asa Gray seemed to have a similar concern.[27]

Another Darwinian subject they discussed was plant geography. Again, their respective interests predated pubication of the *Origin*, but that book provided a new context for the discussions. One example of Watson's skepticism towards some of Hooker's ideas shows that Watson had not lost his interest in the non-British and

---

[25] Yeo 1973 gives the references; see also Watson 1843–47, III, 598; Watson 1870, 201–2, 269. D.E. Allen adds: 'Bentham was notoriously a 'lumper', who, as a producer of monographs on a world scale, preferred a broader interpretation of species than many taxonomists since have found it possible to accept. His *Handbook* of the *British Flora* (1858), which he wrote avowedly as a simplified text for relative beginners, has been credibly criticized as holding back field botany in Britain for several generations.'

[26] See, for example, Guppy 1917, Tutin 1953, and their references.

[27] Watson cited 'Gray's Examination of Darwin & his Reviewers' in a letter to J. Hooker, 5 March 1861, and Watson returned to the subject in his letter of 9 August.

non-Azores aspects of the subject merely because he found it impracticable to publish on them (5 March 1861): 'Shall you extend your glacial area, so as to convey plants from Fernando Ro to Abyssinia? Or, will you construct a ridge of 10,000 feet from one to the other? Are all things possible – with you?' Despite their differences in judgment, however, Watson admired the scope of Hooker's firsthand experience: 'Was there ever a botanist who had seen so much of the earth & its vestments?' And Watson imagined 'how vastly I should have enjoyed the run over Syria with you!'(4 July 1861).

Thus far, their phytogeographical disagreements were well within the bounds of friendly discussion, but Watson's attitude eventually turned personal (7 June [1862?]:

> In consequence of our exchange of letters about your article in Linn. Trans. on the migrations of Arctic plants,[28] I gave up the intention of writing thereon myself ... you alone (so far as I know) have endeavoured to harmonize existent facts of plant distribution in arctic lands with the Forbes-Darwin hypothesis of glacial migration ... the only way in which I could handle the subject now, would be to show you unsuccessful in this attempt; which I must abstain from doing, because you would seemingly regard any such handling as an attack on your Essay, not as a discussion of its subject matter.

It is unlikely that Hooker ignored this unfair assertion, yet his response did not stop Watson's comments. On 11 December [1863?] Watson complained that Hooker and Bentham had unfairly characterized the school of geographical botanists to which Watson belonged as 'fact plodders as distinguished from the imaginative creators', and that if Watson were not too old to do so, he might respond (presumably in print).

Hooker probably asked for specifics, and Watson finally responded that 'the main facts seem to me to show simply a convergence of migration towards the Arctic lands, & do not warrant any conclusion that the species now there originally went thence to the hills of the temperate latitud[e]s & southern hemisphere & then back to the Arctic lands.' Hooker was unable to satisfy him, and on 28 January and

---

[28] J. Hooker's 'Outlines on the Distribution of Arctic Plants' had been read at the Linnean Society on 21 June 1860, and therefore could have been available to Watson in either 1860 or 1861. However, I suspect that he was responding to the published version, which would place his letter in 1862. Some of the earlier letters to which Watson alludes may have disappeared. On 17 February 1861 (in which he responded to a letter, not to the article), Watson wrote: 'I can pretty well understand your uncertainties about species in Arctics. It seems almost impossible for the same person to take large & little views. That minuteness of observation & distinction, of the small-great species splitters, must be long exerted & much restricted; & it is very narrowing to the mind, although a real skill and expertness as far as it goes ...'.

Figure 16 J.D. Hooker (*c*.1870). Photo by Wallich.

1 February 1864 Watson wrote exasperating letters which would surely have ended the relationship had Hooker been less patient than he was. On 28 January:

> Supposing that I should print some animadversions on your application of Forbes-Darwinism to Arctic botany, – can you bear to have the naked truth (that is, what *I* take to be such) sent forth in my brusque style?
> You are so accustomed to eulogy, so little used to plain criticism, that I doubt even your usual good temper being able to bear the latter; unless, indeed, you can take comfort in the thought, 'I am a Hooker: it is only a Watson who criticises me.'

It is ironic that Watson 'lumped' Darwin with Forbes; in reality, Hooker, not Darwin, shared Forbes's propensity to hypothesize land bridges to solve problems of geographical distribution. On 1 February:

> I know that you have never read any treatise on logic or on moral philosophy; otherwise I should have pronounced your ideas of what is 'ungracious' in a scientific discussion as being one of the most remarkable instances of one-sidedness that I have ever seen.

After more gratuitous comments, Watson came to his substantive claim:

> I do accept the glacial hypothesis in a very limited sense, compared with Mr Darwin & yourself. It is a vera causa now in restricted action. But to my judgment, you & D. exaggerate it almost to parody ...

Hooker kept a copy of his own next letter to Watson, giving us a full record of this weird exchange. Patiently, he explained: 'Either I expressed myself unclearly, or you have misapprehended my meaning. I have no wish to stifle scientific discussion, nor to imply that I should take offense if you express dissent.' He expected his Arctic glacial botany to be evaluated on its merits (5[?] February 1864).

Since Watson had apparently challenged Joseph's integrity as well as his scientific judgment, this would have been an appropriate time for Joseph to let the relationship die from neglect. Joseph may have been tempted, but after his father, Sir William Jackson Hooker, died at Kew on 12 August 1865, Joseph wrote a brief biographical sketch on him and sent a copy to Watson.[29] In thanking him, Watson spoke of 'your

---

[29] This 1865 sketch is not listed in Leonard Huxley's bibliography of J. Hooker (Huxley 1918, II, 497–8), but Desmond 1977, 319, lists (without authors) three biographical sketches published during that year, one of which may be by Joseph. Joseph's more substantial biography of his father appeared in 1903.

illustrious Father' and complimented Joseph on having written in excellent taste. Then Watson offered his own candid assessment (12 November 1865):

> Though Sir William made no discovery in science, invented no new method, never converted details into generalization, yet he was a very remarkable man in his own chosen beat, and cannot pass into oblivion.

He followed this brief characterization with a lengthy phrenological assessment, which must have made Watson's response the most unique one which Joseph received from a fellow botanist. In 1866 Watson made an earnest effort to convince Hooker of what he had not been able to convince his father: that the Royal Botanic Gardens, Kew, should make a serious commitment to British as well as to foreign botany. His attempt was prompted not by the death of Sir William, but by the employment at Kew of John G. Baker, who had run the Botanical Exchange Club, but no longer felt he had the time to do so. Quite aside from the need of a center for British botany, there was the question of appropriate use of taxpayers's money (13 January 1866):

> Don't be displeased, if I remark that the great objection to granting so much public money to Kew, lies in the fact that it is granted for the benefit of a very small body of English botanists; 19 in 20 of us (as botanists receiving no scientific advantage from that expenditure). Mind, I don't say that botanical science is not advantaged, & greatly advantaged; but that few of the individuals who take interest in botanical matters, find any advantage to themselves or to their own favorite departments thereof.

However, if Hooker were to commit modest resources to advance British botany, then the political arguments favoring the government support of Kew would be strengthened. Hooker explained that all available funds were already committed, and Watson replied (18 January): 'We look at public expenditure with different eyes, & from different standing points.'

By late 1869, Hooker responded to Watson's plea to do something for British, as opposed to foreign botany, in a way that Watson had not imagined: he decided to write a British Flora for students. His father had written one, which Watson praised in its early editions and condemned in its last edition. Watson advised him on what he thought would make it acceptable (17 December), and when Hooker published *The Student's Flora of the British Islands*, Watson praised it as having hit the medium between Bentham's *Handbook* and Babington's *Manual*: 'Babington places too many very questionable species on equality with the unquestioned', and in avoiding that extreme, Bentham 'too much neglects them, as real though subordinate forms'. Watson thought 'a great step is gained for the rising botanists, by training them through a descriptive Flora to recognize and trace *gradations*', and that *The Student's Flora* would reduce the barriers to acceptance of Darwin's views (30 May 1870).

In return, Hooker sent some corrections to *A Compendium of the Cybele Britannica* (1870), for which Watson thanked him (15 June) and wrote that he could include them in the next supplement (which appeared in 1872).

It was as a professional phrenologist that Watson informed Hooker that he had read Darwin's *Descent of Man* (1871) 'with much disappointment', for 'Darwin is widely out of his fit subject when discussing matters psychological' (1 June 1872).

If Hooker smiled at that judgment, he stopped smiling when he discussed his conflict with Ayrton. Hooker had, as a scientist, dealt successfully with government administrators for all of his adult life, and until 1870 those interactions had been facilitated by a mutual esteem which Hooker assumed would continue. However, Acton Smee Ayrton, who had been born at Kew village in 1816, and who served in Gladstone's administration as First Commissioner of Works, 1869–73, was an unpopular administrator – partly because of his zeal for economizing and partly because of his brusque manner. In Ayrton's attempt to impose his will on Hooker, the government–science issues were minor.[30] Ayrton had an abrasive personality, and the gossip of the day indicated that he had been given the post to keep him from criticizing Gladstone's policies. Hooker shared his exasperation with Watson, who responded with advice. By July 1872 Hooker was on the verge of resigning. Watson was among those who saw that as a serious mistake (letters on 2 and 3 July), and urged him to retreat from his declaration that either Ayrton or Hooker ought to go. Watson reasoned that 'Gladstone likely cares nothing about natural history sciences, and has occasionally evinced a tendency to stick to colleagues through thick and thin.' Watson believed (3 November) that Ayrton had made enough enemies for him to be dismissed during the next session of Parliament, 'so it may be your best bet to put off, not to hasten a crisis. At any rate, put the question rather how far the office-holder (impersonally) not how far Mr. Ayrton (personally) is justified in his interference with you.' Ayrton had not stuck his neck out without cover; Richard Owen was glad to advise him on how to handle Darwin's best friend. That was no surprise; but five years later Watson broadly hinted that he also might have provided some ammunition to the opposition, though not necessarily to Ayrton (5 December 1877):

Some years back, when taxation was more criticized than it has been of late, the estimate [of a favorable] vote for Kew ran good chance to be challenged in the H[ouse] of C[ommons]. One, among other grounds of objection, would have

---

[30] Huxley 1918, II, chap. 35. On Ayrton, see Courtney 1901 and Mac Leod 1974.

been the argument, that *English* taxpayers were not justly called upon to maintain as Herbarium for the service or pleasure of *Foreigners*, chiefly. Perhaps the Kew Annual Reports may have alluded rather too much to the visits of Foreign botanists, without showing usfulness to those of Britain specially.

The occasion for this reminiscence was Hooker's negotiation to obtain Watson's herbarium for the Royal Botanic Gardens upon his death. Watson foresaw that 'allotting a space in Herb[ariu]m especially for the service or pleasure of purely British Botanists will be an obvious answer to any such line of anti-[science] argument, if tax-questioning should again become popular in H. of C.'

A Protégé for Watson

As a botanist with no institutional connections except membership in the Linnean Society, with a combative personality and reclusive habits, Watson was in a poor situation to attract a protégé. On the other hand, Watson could reach out with generosity if some worthy person had both deference and an obvious need. John Gilbert Baker (1834–1920)[31] had both, and Watson's generous response allowed the relationship to turn gradually into one of patron and protégé.

Baker grew up in Thirsk, Yorkshire. While attending a Friends' School in 1846, he began collecting plants, and in the next year he became curator of the herbarium of the Bootham School Natural History Society (the oldest such society in Britain). In 1849 he sent the first of many notes and articles to the *Phytologist*, and these were so highly regarded that Edward Newman attempted unsuccessfully to recruit him as editor after Luxford died in 1854. Baker's schooling led not to college, but to his father's drapery and grocery business. He remained a hobbyist who published regional notes on plants gathered by himself or by other members of the Thirsk Natural History Society.

Two developments propelled him from a provincial to a national sphere of activity. First was the resolution passed by that Society in November 1857, under his presidency, to run a Thirsk Botanical Exchange Club (TBEC). This action coincided with the collapse of the London Botanical Society, its exchange club. Under his able curatorship (1857–68) and secretaryship, (1868–79), TBEC attracted a national membership, which prompted Watson to transfer to it LBS's records and the remaining stock of its Proceedings. Watson was familiar with Baker's competency, not only from his writings in the *Phytologist*, but also from his separate publications:

---

[31] Allen 1986, chap. 6 et *passim*; Britten 1920; Desmond 1977, 30; Freeman 1980, 37; Stafleu & Cowan 1976–92, I, 104–5.

A Supplement to Baines's Flora of Yorkshire (1854, coauthored by John Nowell) and *The Flowering Plants and Ferns of Great Britain: An Attempt to Classify Them According to Their Geognostic Relations* (1855). Both works showed Baker's interest in plant geography. In 1863 he published an expanded and revised edition of what he had coauthored in 1854, *North Yorkshire: Studies of its Botany*. These worthy efforts led to a pleasant, if distant, mutual admiration between Baker and Watson. The second development in Baker's progress from provincial to metropolitan botanist was a catastrophic fire in May 1864, which destroyed not only Baker's house and shop, but also his botanical library, notes, herbarium and copies of his most recent book.

By then, Watson loomed large among Baker's botanical colleagues, as he remembered much later:

> when a botanist of a younger generation, who was all the while looking up to him as a great master of science, would venture to call uninvited, he would, after devoting several hours of his time to instructing and entertaining him, thank him on parting for the visit, in a way to which it was very difficult on the spur of the moment to make any suitable response, because it was so unexpected. The encouragement and help which he gave to many young beginners in Botany is remembered by them with feelings of gratitude and affection ... when I left school and became immersed in business ... I should probably have drifted away from [botany] altogether if I had not made his acquaintance. In answering letters, in helping his correspondents to name critical plants and fill up gaps in their herbaria, in referring back to his notes and catalogues (as he was asked to do so often in his latter years), to explain the details on the faith of which he had stated some record in general terms, his patience and assiduity were unbounded, and no one but those who lived with him knew what a large proportion of his time was often occupied in this way. In helping by money to the full extent of his pecuniary means, and by judicious counsel, such of his friends and acquaintance as fell into illness and misfortune, his liberality was very great.[32]

Baker must have known that Watson was the effective organizer and a generous contributor to the rescue fund collected among botanists after the fire. One of the letters which Watson wrote on Baker's behalf has survived; it shows Watson's organizing propensity put to good use:

> A movement has begun in London towards a subscription fund to aid in replacing the library, but in a small and private way only, which is little likely to be of much real service.

---

[32] Baker 1883, xx–xxi.

I have requested ... Botanists whose names would carry weight, and serve to show a more widely extended sympathy, will unite in the matter, by letting their names appear in a printed Circular, either as approving & sanctioning the project, or as being members of a Committee formed with that object.[33]

A grateful Baker later wrote to the subscribers that the fund was 'far more than sufficient to replace all my botanical belongings which money can restore'.[34]

In 1866 Joseph Hooker hired Baker as First Assistant at the Kew Herbarium. He retained that position until he became Keeper of the Herbarium in 1890. Before he went there, Watson sent him some unwise advice: 'take your own convenience when you go to Kew; that is, suit the date to your own needs, not to his. Unless you begin by showing that there is another's ties to be considered, and not his own wishes solely, you may become a slave more than is fair or agreeable.'[35] Watson hoped that Baker would be able to bring the exchange club to Kew and also create there 'a general Centre of *British* Botany,' but upon proposing these ideas to Hooker, Watson found that he refused to change the priorities set by his father.[36] Watson then complained to Baker about his boss:

[Hooker's] reply is not at all unfriendly, but still to the effect that not a minute of your (official) time could be spared for any such purpose ... he holds British botany in a sort of contempt, & being very much occupied with his own papers & publications, he naturally wishes you & Mr. [Daniel] Oliver to take as much off his hands as possible, of the Kew duties.[37]

At age 32, Baker was too experienced to be swayed by such spleen. He performed his assigned duties in foreign botany without complaint, and found time on the side for British botany, just as he had when helping his father run the family business. However, he soon relinquished the curatorship of the exchange club in favor of the much less onerous position of secretary.

Baker published articles on phytogeography in *The Gardeners' Chronicle* which he later collected in *Elementary Lessons in Botanical Geography* (1875). He relied heavily on Watson's works for details on Britain and on Alphonse de Candolle's *Géographie botanique raisonnée* for details on the rest of the world. He sent a copy to Watson, who thanked him and raised several questions, for example (28 September [1875]):

---

[33] Watson to Balfour, 27 May 1864.
[34] Quoted in Britten 1920, 234.
[35] Watson to Baker, 19 January 1866; the whole letter is published in Lousley 1962.
[36] Watson to J. Hooker, 13 January 1866; Allen 1986, 75. For Watson's identical suggestion to William Hooker, see chap. 7.
[37] Watson to Baker, 19 January 1866.

Figure 17 John Gilbert Baker at work.

Page 110. How is it to be proved that 'each species has originated from a single centre'? Can you prove the 'origin' of any single species and where that origin was? I know that is the prevailing or generally accepted idea; but it is a mere guess ...

For his part, Watson continued to collect phytogeographical data on British plants and to refine the precision of his accounts of species distributions. After *A Compendium* (1868–70) of more than 650 pages came *Topographical Botany* (1873–74), with 740 pages. He was working on a second edition of the latter when he died; Baker and W.W. Newbould completed and published it in 1883.[38]

Baker was the executor and chief beneficiary of Watson's final will, receiving the house and land at Thames Ditton and Watson's books and botanical collections. He had persuaded Watson not to burn his herbarium,[39] but all the letters did go up in smoke. Watson explained to James Britten (7 April 1878):

After hurriedly writing the 'Topog' Bot^y', while the data still existed, I burnt (inter alia) a wheelbarrow full of botanical correspondence, misc. localities, &c. This was done in order to spare my Executor the tedious trouble, which would have been otherwise entailed upon him, in having to look through such a mass of letters & papers in order to select such as might seem to him needful to keep a while longer.

Watson's Other Relationships and Influences

A few other hardy souls besides Baker avoided Watson's barbs and followed his lead. The London naturalist Alexander Goodman More, enthusiastically began his serious study of botany in 1852:

His purchase this year of Watson's *Cybele Britannica* had doubtless much effect in influencing his mind in this direction, for no other book so completely dominated his whole line of thought, and throughout life, both as botanist and zoologist, he was the most ardent of 'Cybelizers'.[40]

---

[38] 'The ledger-size books in which vice-county records were entered are now at Kew, but a set of companion records are in the Botany Library at the Natural History Museum.' – D.E. Allen.

[39] 'Today it stands in an honoured place, halfway up the stairs, in the Herbarium at Kew – still in Watson's cupboard.' – D.E. Allen.

[40] More 1898, 26. On A.G. More (1830–95), see Desmond 1977, 449; Praeger 1949, 134–5; Stafleu & Cowan 1976–92, III, 575–6.

More and a collaborator used Watson's system of 18 British provinces and 38 subprovinces to describe the distributions of British butterflies (1858). That publication led Watson to invite him to spend a few days at Thames Ditton. Watson then allowed More to republish the 1852 map of provinces and subprovinces in a much longer study on the distribution of British birds (1865).[41]

Had Watson obtained the botany chair at an Irish university in the 1840s, he would surely have extended the scope of *Cybele Britannica* to include Ireland, but since he had not, others took up the challenge. More moved to Ireland, where he collaborated with Scotsman David Moore (born David Muir), curator of the Royal Dublin Society's Botanic Gardens, on *Cybele Hibernica* (1866).[42] Although the older Moore was the senior author, its completion owed much to the younger More's enthusiasm and to the example of Watson's work. Like Watson's *Cybele*, theirs received a favorable review from Alphonse de Candolle.[43] Moore and More summarized some of their findings at the International Horticultural Exhibition and Botanical Congress at London in May 1866; they went beyond Watson's mapmaking to show isotherms in relation to the distribution of some species in Ireland.[44]

Watson had been sufficiently impressed by More to recommend him to Darwin as a reliable correspondent. Furthermore, when More applied for an Irish chair of botany in 1869, Watson wrote a letter for him, attesting to:

> his high attainments in [botany], which have been long known to me through his published writings as well as by frequent correspondence and personal acquaintanceship. Mr. More has acquired a critically exact knowledge of British plants, which is fortunately also combined with a wider knowledge of the science of botany in its more philosophical branches. His special attention to the botany of Ireland is amply shown by the *Cybele Hibernica*, and by his discovery of plants there which had previously remained unknown to the botanists of Ireland.

However, after collecting and printing the recommendations from an impressive group of botanists, More decided he would 'not be equal to speaking in public', and withdrew his application.[45] His actions echo Watson's own behavior during the 1840s, and also in another respect: More began preparing an enlarged second edition of *Cybele Hibernica*, which was completed after his death by Nathaniel

---

[41] Boyd & More 1858 and More 1865.
[42] On David Moore (1807–79), see Desmond 1977, 447; Praeger 1949, 133–4; Stafleu & Cowan 1976–92, III, 568–9.
[43] Candolle 1869; a portion of it is reprinted in French in More 1898, 232.
[44] Moore & More 1867. Their map is discussed and reproduced in Nelson 1993a. See also Desmond 1977, 447 and 449; Praeger 1949, 133–5.
[45] More 1898, 103, 130, 139, 151, 153, 229 and 232.

Colgan and Reginald W. Scully, just as Baker and Newbould had completed the second edition of Watson's *Topographical Botany*.[46]

An anonymous essay, 'Botanical Geography', appeared in the *North British Review* in 1854; ostensibly, it reviewed *Cybele Britannica*, volumes I–III. It was written by Rev. John Fleming, a zoologist and geologist who had become Professor of Natural Science at New College, Edinburgh. It is impressive that a Scottish magazine was willing to devote 22 pages to this slightly esoteric subject. To prepare for the task, Fleming read a few works on plant geography; yet he was still incompetent to evaluate Watson's volumes. What he wrote was bombast, interspersed with lengthy quotations. He judged *Cybele* as being 'scarcely more than a mechanical operation, conducted, however, by one to whom the subject was familiar in all its bearings'.[47] Fleming performed his Christian duty of being a peacemaker by mentioning Forbes's praise of Watson and remaining silent about Watson's attack on Forbes. However, Forbes's plant geography did not escape Fleming's criticisms any more than Watson's did. If Watson read this, he did not think it worthy of response.

Since John Lindley was horticultural editor of *The Gardeners' Chronicle*, his authorship of the anonymous attack on Watson and *Cybele* volume IV would have been obvious to many botanists. Watson's response came in *Part First of a Supplement to the Cybele Britannica*, which did not have nearly as broad a readership as *The Gardeners' Chronicle*. Watson defended his candid pronouncements that sometimes caused offense, and claimed that he had not bothered to read the review because his two correspondents who mentioned it indicated that it was an uncritical diatribe. That characterization was not unfair, yet the review interests us because it shows how one, and most likely others, responded to Watson's caustic remarks:

> We are sorry not to be able to appreciate the merits of this Cybele Britannica. And before we dismiss it we must draw attention to the extraordinary language which its author employs when speaking of others. Poor Forbes he calls blundering and false, a plagiarist, a man of reckless hardihood of assertion. Gentlemen who hold botanical chairs are accused of filling them for the mere sake of putting money into their pockets, botany being 'useless to medical practitioners'; an assertion which shows that Mr. Watson really knows very little about the subject on which he has undertaken to write. It is announced that very

---

[46] More, Colgan & Scully 1898. On Colgan (1851–1919) and Scully (1858–1935), see Desmond 1977, 141 and 549; Praeger 1949, 66 and 153.

[47] [Fleming] 1854, 515–16. On John Fleming (1785–1857), see Desmond 1977, 225–6; Page 1972; Rehbock 1983, 120–3; Rehbock 1985.

few English botanists (?) seek more than a petty amateur knowledge of species. We are informed that the prevailing peculiarity of the botanical mind is not always reasoning with strict accuracy and soundness; M. Alphonse De Candolle cannot reason on 'causation and dependence'; and so on. If Mr. Hewett Watson can condescend to take advice he will employ another tone hereafter.[48]

However widespread Lindley's sentiments were, Watson was not inhibited from attempting to start a new journal of British botany in 1861. He detested the *Phytologist* under Alexander Irvine's editorship and hoped to steal its readers. He recruited the mild-mannered J.T.I. Boswell (*né* Syme, afterwards Boswell-Syme) as editor ('a good botanist, but unfortunately not a suggestive man'), Edward Newman as publisher, and presented the plan as a group effort in his circular and his letters to prominent botanists asking for their support.[49] With tongue in cheek he reported to Trevelyan (1 March 1861):

I have alarmed Prof. Arnott, by telling him the 'species-splitting' is sure to be represented, since some of the active promoters (A.G. More & J.G. Baker) are addicted thereto. Equally I dare say, Dr. J.E. Gray is alarmed at lending his name to a journal, where I tell him Darwinism is quite likely to creep in.

'The British Botanist' never appeared, but the above substantive issues were not the reasons. Rather, many of those who agreed to list their names as supporters nevertheless did not relish the idea of Watson having a heavy editorial influence.

Politicians know that you cannot beat someone with no one, and the same maxim applies to periodicals. Berthold Carl Seemann, who, with a brother, had edited *Bonplandia: Zeitschrift für die gesammte Botanik* from 1853 to 1862, moved from Hanover to the London region and began editing the *Journal of Botany, British and Foreign* in 1863.[50] Even with his close ties to Kew, it would have been quite a feat for a peripatetic foreigner to immigrate to a country where a prominent native son was trying to establish a scientific journal and without any local sponsorship beat

---

[48] Lindley 1859; the enclosed question mark is Lindley's. On Lindley, see references on page 190. n. 51. For Watson's response and criticisms of Lindley's botanical judgment, see Watson 1860, 4–8 and 27–9.

[49] Watson's assessment of Syme is in a letter to Trevelyan (1 March 1861). His first circular was dated 1 December 1860 and is included with letters of solicitation to J. Hooker (13 February 1861) and to Trevelyan and to George Bentham (both on 25 February 1861). He stated to each of them that a longer prospectus would be issued with the names of botanists who supported the need of the journal. The story is told with a somewhat different interpretation in Allen 1985, 71–2. On Boswell and Newman, see Desmond 1977, 78 and 463.

[50] On Seemann (1825–71), see Boulger 1885 and Desmond 1977, 550.

him to the draw. The initiative probably came from Hooker, who succeeded in remaining out of sight. Watson accepted this journal as meeting his needs (the *Phytologist* did fail) and began publishing in it the following year. Around 1869, the restless Seemann decided to manage tropical properties in Nicaragua and Panama; he died of a tropical disease in 1871. Therefore, Baker and Henry Trimen were recruited as assistant editors.

Although Watson moderated his previous propensity to attack others without provocation, he still defended himself when provoked. Two examples show that it still took very little to provoke him. In 1863 George Bentham, in his presidential address to The Linnean Society, referred to progress in phytogeography without mentioning Watson's work.[51] As soon as Watson read the address, he sent Bentham a challenge (23 December 1863):

> ... you say that Botanists are *now* aware that geographical botany does not consist *merely* in showing the boundaries of a species in one *minute* portion of its area. This would seem to imply that writers on geog¹· botany did lately hold that absurdly narrow view of their study.

Watson complained that botanists were assuming Bentham meant that *Cybele Britannica* was useless, despite Bentham's reliance on its distributional data in his *Handbook of the British Flora* (1858). Bentham assured Watson that he meant no aspersions and regretted the misunderstanding. Watson then thanked him (25 December 1863) for his 'courteous & satisfactory reply', but the earlier negative experience rather than Bentham's later attempt to calm troubled waters persisted in Watson's memory. Subsequently, Baker lent Watson a pamphlet by Bentham which Watson admitted to reading only cursorily before reaching a conclusion (28 March ?):

> It strikes me as bordering too closely upon the verbiage of senile egotism. The writer appears to think that the world, – the small world of technical botanists, – will be unable to progress after his decease unless he instructs them what to do & how to do it.
>
> Now, for a man who has never shown any originality of thought, discovered nothing, invented nothing, but has always been an imitative plodder, rather behind than in advance, – for such a man to fancy that he is the one to *guide the future*, does to me appear a curious sample of egotistic mis-understanding of himself. The cast of Bentham's mind is Chinese more than American.

---

[51] On Bentham (1800–84) see Bentham 1997 and Desmond 1977, 58–9; Jackson 1906 mentions this episode on 197; Bentham responded to Watson's letter on 24 December; Stafleu & Cowan 1976–92, I, 173–81; Taylor 1970.

The second provocation was Henry Trimen's anonymous three-part review of *A Compendium of the Cybele Britannica for the Journal of Botany*.[52] Watson did not respond to the first two reviews, but published a reply to the third and addressed it to Baker for transmission to his fellow assistant editor, the presumed reviewer. Trimen replied directly to Watson, who then published a response to Trimen's letter. The thrust of Watson's retorts was that Trimen was too inexperienced to be a competent critic of Watson's work. Since Trimen's remarks were mostly favorable, and since the readers of the *Journal of Botany* were mostly familiar with Watson's works already, Watson's self-defense seems unnecessary.

Watson's esteem for de Candolle encouraged him to think that he might share Watson's low opinion of William Hooker's intellect. He had sent de Candolle a copy of *Cybele Britannica*, volume III (1852), which contained his attack on the sixth edition of Hooker and Arnott's *British Flora*. Next, on 4 September 1859, when thanking de Candolle for his favorable review of *Cybele Britannica*, Watson responded to some comment in that review or in a lost letter that accompanied it by criticizing Hooker:

> technical botanists are usually bad reasoners. Take Sir William Hooker in example. His talent in botany, you are well aware, is depictive and descriptive. He represents and describes species or genera. You will admit that he has done good service to botany. And (making some allowance for haste and quantity) that he has been a good depicter and describer. What more is he? Nothing. Utterly unable to reason on any subject, whether botanical or not. He has not a capacity to understand your *Geographie Botanique*. And if anybody attempts to converse with him on any subject requiring the faculty of *reasoning*, he is found to be below mediocrity – hardly better than an idiot.

Despite such testimony (all of which is not quoted), de Candolle not only published a highly favorable essay on Hooker's life and work (1866), but also sent a copy to Watson. The latter pleaded poor health in taking three years to respond, then partly agreed with de Candolle's evaluation of Hooker while adding his own alternative assessment (August 1869):

> Certainly he did a vast amount of work, and gave a great impulse to botanical exploration abroad. You could not fail to know (but probably did not wish to speak of) one remarkable peculiarity in him, id est, *the very narrow quality of his intellect*. Compare him with your Father in that aspect. The latter was a Describer of plants, like Sir W. Hooker, but he was also much more than a simple Describer

---

[52] Trimen 1868–70. On Trimen (1843–96), see Desmond 1977, 618.

of species and genera. Hooker was a Describer and Depicter, and nothing more. He was incapable of methodizing or classifying details – incapable of connecting ideas in the way of cause and effect – incapable of bringing out any third idea or conclusion from two or more simple facts brought into relation with each other – in short, utterly incapable of ratiocination. It was impossible to converse with him on any matter of mental science or moral philosophy. In describing and depicting plants, he was a genius; in almost all else, in all that required thought and reason, he was much below mediocrity. He was not a '*Homo sapiens*', but nearer to a *Simia intellectualis*. He had the curiosity of a monkey in finding and examining objects: that was his genius, with a training of it for science.

It never occurred to Watson that de Candolle might view such an unfavorable assessment as a slap at his own ability to judge other scientists. Watson ended his letter regretting that Joseph Hooker could not accept criticism with good humor. Watson was glad that de Candolle was not 'one-sided' like Joseph Hooker.

De Candolle sent thanks to Watson as he received each printed part of the *Compendium*, and Watson responded (30 January 1871) that he hoped to 'see another volume of the Prodromus, but my life should now be nearer its end, at 66–7 years'. Watson outlived this pessimistic expectation and on 31 January 1874 wrote to thank de Candolle for sending him the 'Prodromi ... Historia, Nurmeri Conclusio'. In return, he sent the First Part of his *Topographical Botany*, and would send the Second Part whenever printed. He lamented that it was data without interpretation:

> It was my wish and expectation through [the] years, to write a work in criticizing review of the various hypotheses suggested to account for actual distribution – not present distribution, as in accord with present climate, but as to be accounted for historically & geologically. The question seems to me to be triple:
> 1 What is present distribution?
> 2 What was past distribution?
> 3 *How* has no.1 evolved from no.2?

He then complained:

> As to the '*How*', I am dissatisfied with anything hitherto set forth. Darwin, Forbes, Hooker, in this country, may each have a glimmer of truth, but each selects a few facts in support and ignores the many facts in contradiction. They do not explain [the] whole [of] nature; they only argue for one-sided views. They fit the facts to the hypotheses; do not show a theory evolved from facts.

Somewhat inaccurately, Watson imagined that 'the natural taste with me is really more towards ratiocination, than towards observation,' and he hoped that his 'humble books' would aid 'some future botanist in attempting the higher purpose' of interpreting his data. De Candolle was more optimistic about what Watson had achieved, and published a final tribute to his lifetime contributions to plant geography in 1880: Watson's works were 'un vrai modèle de persévérance, de méthode et de précision, avec des notes ou chapitres additionnels d'une grande portée philosophique'.[53]

Watson's correspondence with Babington continued regularly throughout 1860 and on until 6 May 1861 (nine letters). On 13 May 1859, Watson had speculated on the potential impact of Darwin's then still unpublished *Origin* on systematic botany, but only one of these nine letters resumed that discussion after the *Origin* appeared. On 6 March 1861, Watson recommended to Babington three essays by Asa Gray defending Darwin's theory. Perhaps Watson felt that Babington might be receptive to a defense of it written by a fellow Christian. Neither Watson's last extant letter to Babington nor Babington's letters to Balfour indicate why the Watson–Babington correspondence ended in May 1861. Possibly Babington grew weary of Watson and decided to write only to colleagues from whom he derived some enjoyment. Babington was not so annoyed with Watson in 1864 to refrain from praising his studies on primula hybrids.[54] However, it is revealing that Babington waited to join the Botanical Exchange Club – which should have been as important to him as it was to Watson – until immediately after Watson died.[55]

Babington's feelings toward Watson were shared by Balfour. In 1864 they twice exchanged letters on a humanitarian matter and did not confront the reason they had drifted apart. Watson was an 'injustice collector', as he himself knew, and it was not until 8 February 1871 that he found an occasion to summarize the cause of the rift, as best he remembered, mentioning complaints not previously aired:

> Assuredly, I left Edinburgh in 1833 with feelings of the utmost cordiality towards yourself, Brand, [William Hunter] Campbell, & others; & the like feelings continued for some time, until I felt myself unhandsomely treated, and my proffered co-operation with B.S.E. rejected, not simply by non-acceptance (which I fully admit was altogether optional with you) but by the rudeness of

---

[53] Candolle 1880, 177–8; see also 350–1. I am endebted to Jean-Marc Drouin for locating and transmitting to me the entire commentary from which this quotation is taken. This passage is possibly the basis of the claim: 'When M. Alphonse De Candolle lately made out a list of botanical epochs, he counted the publication of the Cybele as one of them'; Baker 1883, xix.
[54] Babington 1864.
[55] Allen 1986, 78.

utter disregard – of non-notice altogether. Then I joined the London B. Society instead, and the whole course of my subsequent botanical life & doing was more or less altered.

I may have been right or wrong in the conjecture, but I gave you all [in Edinburgh] credit for being influenced by C.C. Babington, & adversely to me.

Balfour claimed no knowledge of the incident, and Watson therefore reconstructed the circumstances to the extent that his memory allowed (17 February 1871):

When the B S Ed was started, I became a member & recommended others to do so. I recollect going one evening to B S London especially to advise that its members should adopt the Edinb. Catalogue as their standard of nomenclature, &c. if they resolved to keep up an independent society. My idea then was that the Edinburgh Socy. should be held the British one, [with] the London Socy. [being] a sort of local offshoot.

I cannot at all recollect the precise year, but I wrote to Mr. Brand a sort of proposal or proposition nearly to this effect: That I was much inclined to keep up a close intimacy with Edinburgh & Scotland, partly botanical, partly phrenological. To this end, if members of B.S.E. thought fit, I would make arrangements for remaining some weeks of each year in Edinburgh, assisting in the exchanges or distributions of the Society, and otherwise [attending to BSE business]. To this proposal I never received any reply in any form.

Someone without a contentious personality might then have written to some other officer of the Society to find out whether his offer had been mislaid and what the Society's response was. But self-centered as Watson was, 'The neglect seemed, in my reading of it, to border on the contemptuous.' Even in 1871 he only imagined that Balfour had forgotten the incident – not realizing that it may never have come to his attention at all. However, in retrospect, he could admit that he might share the blame for the situation. When he felt slighted, his strong inclination was: 'Smite back.'

To Baker, Watson admitted having been too harsh in criticizing fellow botanists. Baker recalled in his warm obituary on Watson:

One of his pet prejudices was an objection to new names for plants, and the name-givers were a favourite target for his arrows. And thus it comes that those old papers in the 'Phytologist' and the tail-paragraphs in the geographical books are often full of lively personal interest. 'Ah!' he said more than once during the latter years of his life, as we talked over these things on the quiet Sunday afternoons, 'they read too bitter. You don't know what it is to have a large organ of destructiveness.'

In other words, Watson used phrenological notions to blame his combativeness on heredity. Occasionally, he refrained from assigning blame when botanical judgments differed. This was easiest when his disagreement was with Baker, as seen in a letter he wrote to James E. Bagnall (12 March 1878):

> Thank you for 'The Distribution of the Genus Rosa in Warwickshire'.
> 'Each to his taste.' Mine is certainly not that of looking at Dogroses comparingly with the names and descriptions of my good Friend J.G. Baker. Indeed, I doubt whether I can put correctly (at any rate, not put confidently) any one of his thirty varietal names of canina to any living bush in Surrey; – unless, indeed, I except a bush here and there, pointed out by himself as this or that variety, and his applied name for which I may chance to recollect, not by the technical characters of the bush but by its place in the hedge, &c.

Whether or not he ever cared about the varieties of roses, at age 74 life seemed too short to worry about sorting them out. On the other hand, he had not lost interest in British botany as a whole. More than a year later (30 December 1879) he thanked Bagnall for his seasonal good wishes, admitted growing weaker 'in brain & muscles,' but still mentioned the latest botanical news:

> the re-discovery of *Scirpus parvulus* on the coast of Hants, in a new locality, not in the old one; which latter, through many years, was the sole ground for keeping this Scirpus in our Floras as an English (extinct) species. Then came its discovery in Ireland, followed by Dorset & Devon, and re-discovery in Hants in 1879. – This is an illustration of the increased attention bestowed on local botany of late years.

A Liverpool merchant, Thomas Radcliffe Comber, who was born in Brazil and spent five years in India, was also inspired by *Cybele Britannica* to make wide-ranging phytogeographical studies. The broad geographical scope of his articles was in sharp contrast to the provincial journal in which they appeared, though one of them was reprinted in the *Journal of Botany*.[56] More prosaic, but also showing the importance of Watson's work, was B. Daydon Jackson's compilation of 'Local Catalogues Used in Preparing Watson's *Topographical Botany*' (1883). It was based on Watson's documentation for the second edition which Baker and Newbould had recently prepared for publication (1883). Jackson saw the need for his compilation when he began using the second edition.[57]

---

[56] Comber 1873, 1874 and 1875; the last being reprinted in 1877. On Comber (1837–1902), see Anon. 1902 and Desmond 1977, 144.

[57] On Jackson (1846–1927) see Desmond 1977, 339; Stafleu & Cowan 1976–92, II, 396–9.

Although *Topographical Botany* contained the most precise data assembled up to that time on the distributions of British plants, it was not 'user-friendly'. Watson subdivided Britain first into six zones, then into 18 provinces, 38 subprovinces, and finally into 112 counties and vice-counties. The scheme is too complex to keep in one's mind unless one uses it constantly, and it has been supplemented in this century with small maps for each species. However, since Floras continue to be written of counties, and since county boundaries have subsequently proven liable to frequent modification, the Watsonian vice-county system is still in use because of the greater consistency it provides.[58] British botanists partly follow Watson's terminology. The terms 'plant geography' and 'botanical geography' declined in usage in favor of his 'phytogeography'. His objection to the word 'station' for what is now called 'habitat' may have reduced the usage of the former word in English, though his alternative, 'situation', did not gain currency. (By 1870 he had given up 'situation' for 'habitat'.) Neither did his term 'cybele' survive (for a catalog of plants 'to show their relations to the earth, as local productions of the ground and climate'), possibly because few other botanists wrote books on the distributions of particular species within a country. He originated the British convention of capitalizing 'Flora' to designate works on the plants of a region and using the decapitalized 'flora' to refer to the actual plant species of a region. He explained the difference between 'flora' ('totality of the species in any given country') and 'vegetation' ('total mass of its individual plants').[59]

Watson's influence was also conveyed during his lifetime and afterwards by his activities in botanical exchange clubs. He had originated the idea in 1830, others first implemented it in 1835, a few years later he introduced the scientific rigor which rescued its usefulness, BSL's exchange club was the Society's only activity that Watson really supported – until BSL folded. In 1857 Baker stepped into the breach and ran TBEC until his home and business burned in 1864. He and the exchange club, down to 23 members in 1866, moved to London and struggled on into the next century. In December 1884 a breakaway group took the name of the Watson Botanical Exchange Club, and it lasted for forty years. The Botanical Society of the British Isles developed out of the exchange club which the BSL had originated, gradually becoming larger and more substantial after 1900, and it acquired its present stature and new name after the second World War. *The Journal of Botany* became a

---

[58] Allen 1986, 153–8; Dandy 1951 and 1969; Druce 1932; Heath & Perring 1975; Lousley 1951; Nelson 1993a, see references cited on 391 and listed on 402–3; Walters 1957.

[59] He wrote a hyphenated 'phyto-geography'; see especially Watson 1847–59, IV, 372; discussion of 'station' is on 96; suggestion about when to capitalize 'Flora' is on 385; distinction between 'flora' and 'vegetation' is on 412. Cybele was the name of a Greek goddess who presided over the earth's productions; Watson 1847–59, I, 2. For his use of 'habitat', see Watson 1870, 114.

war casualty in 1942. The new Society voted to name its new journal *Watsonia*, 'in commemoration of the man whose particular set of interests the Society faithfully embodied and to whom it substantially owed its very survival'.[60]

There is an oil portrait of Watson painted in his later years that shows in his facial lines, hanging jaw and stooped posture some of his pains in old age.[61] On 1 October 1873 he wrote to Trevelyan that he no longer felt able to spend a night away from home, but that he still took day excursions on the train and would wander away from a railway station to collect plants for a few hours before returning home. Four years later, he was no longer impressed by the convenience of the railroad, but rather the reverse (as he confided in Hooker, 5 December 1877): 'No one but the sufferer can truly realize the ever recurring inconvenience of a stiff joint in travel. I cannot go even from Ditton to London without feeling the inconvenience of a Rail carriage ...' In December 1880 he fell and injured his left foot and ended his botanical work, because (he explained to George Claridge Druce, 28 May 1881) he had constantly taken opiates for pain ever since.

Grace Eastmond, Watson's 'valued Housekeeper', answered Hooker's last letter to him ([14 July 1881]):

I am truly grateful for your kind note received this morning.

It is with great grief I am obliged to inform you that Mr. Watson is too seriously ill to receive visitors. He now keeps his bed, refuses food, & seems utterly unconscious of all around him.

From her signature, Hooker saw the correct spelling of Eastmond, which Watson never knew. She was with Watson when he died at 8:30 a.m. on 27 July 1881. She received in his will £100 and:

all my household stores, furniture, ornaments, tools, clothes, watch, and other objects in and about my dwellinghouse and gardens, a considerable portion among such articles being truly her own personal property, although used in mingled connexion with my own.

---

[60] The quoted phrase is in Allen's words: Allen 1986, 140–1; he tells the complex history from the collapse of the 1850s down to the organization of BSBI after the Second World War in chaps 5–11.

[61] It was probably painted at Baker's request, and there is a photograph of Watson taken at about the same time – probably to aid the anonymous artist. After Baker died, the portrait came into the possession of Mrs Louisa S. Harvey, who presented it on 2 July 1921 to the Royal Botanic Gardens, Kew, and it now hangs in a hall of the herbarium on the second floor; *Kew Bull.* (1921, 345). Baker also gave to the Royal Botanic Gardens Watson's herbarium, 49 of his books, some pamphlets, and Watson's local flora lists, which are kept near the portrait.

Eastmond may have had another residence to which she took her goods, or she may have continued living in Fern Cottage.

As the executor of Watson's estate, Baker probably ordered the printed announcements of Watson's death, and he added handwritten notes at the bottom, inviting people to the funeral that was held on 3 August at 3 p.m. To Hooker, he wrote: 'I shall be very much obliged & I am sure Mrs Eastmond[62] & the relatives will be very much pleased if you can make it convenient to come to the funeral.' In Watson's obituary, Baker was well within Victorian tradition when he omitted mention of Eastmond – even though she had spent more time with Watson than had any other person – but he did state which botanists attended Watson's funeral: himself, Thomas Bates Blow, George E.S. Boulger, James Britten, Dyer H. Groves, Joseph Hooker, B. Daydon Jackson, M.A. Lawson, W.W. Newbould and J.L. Warren. Their presence was implicit testimony that Watson's contributions to botany outweighed his personality flaws. He is buried in the cemetery beside the venerable St Nicholas Anglican Church in Thames Ditton under a full-length polished tombstone on which is inscribed a Biblical quotation in recognition of Watson's lifelong generosity with his wealth: 'It is more blessed to give than to receive' (Acts 20:35). Most likely, the verse and tombstone were ordered and paid for by Watson's chief beneficiary and executor, Baker.

Darwinian Parallels and Contrasts

Darwin was already a prominent zoologist and geologist before he published *On the Origin of Species*, but in 1860 he stood in the first ranks of world science. Many scientists challenged his leadership, but were unable to discredit his achievement. He had opened the door for generations of scientists, and his subsequent researches led the way. He was gratified by his successes, but discomforted by some of the opposition which his theory met. He suffered more from ill health in his later years than did Watson.

Darwin had the enthusiastic support of Huxley, Hooker and many other younger scientists in defending his theory of evolution in the public forum, but he led the way in publishing further researches to support his theory.[63] Until the mid-1850s, his researches had divided between zoology and geology, but from the mid-1850s onwards he divided his researches between zoology and botany. This shift was encouraged by his close interaction with a superb botanist, Joseph Hooker. The good

---

[62] 'It was normal 19th-century practice to call unmarried servants getting on in years "Mrs."' – D.E. Allen.

[63] Bowler 1990, chap. 8; Irvine 1955.

relationship in the 1850s and early 1860s between Darwin and Watson did not last. Nevertheless, Darwin continued to use Watson's publications in writing his books.

Darwin was not corrupted by fame. He was pleased with his successes and accolades, but the very private life he had established at Down House in the 1840s continued. It was reinforced both by poor health and by the hostility which his theory aroused. His later years were even more scientifically productive than Watson's, and he was also involved in raising his children, maintaining a wide-ranging correspondence, supporting local charities and other causes. His health declined more rapidly than Watson's. Watson's funeral had attracted a respectable showing of botanists; Darwin's, by contrast, was a national event, with burial in Westminster Abbey.[64]

Watson was far from being the only evolutionist who retained doubts about the adequacy of Darwin's theory; even at the time of Darwin's death it was unclear to many scientists how much of his theory could withstand long-term scrutiny.[65] Only in the twentieth century, after evolution and genetics were reconciled, did it become possible to see that the main features of Darwin's theory retained their validity. Darwin's achievements never faded from memory as Watson's did, and today those achievements are better understood and appreciated than ever before.

---

[64] Desmond & Moore 1991, chap. 44; Moore 1982.
[65] Bowler 1989, chaps 7–11; Clark 1984, chaps 7–16; Desmond & Moore 1991, chaps 33–44; Ruse 1979, chap. 8; Vorzimmer 1970.

Conclusions

Hewett Cottrell Watson was an important member of the British scientific community during the Victorian period. There are striking similarities and differences between him and his colleague, Charles Darwin. Since they had similar backgrounds and interests and were diligent researchers throughout their lives, why did Darwin have a much stronger impact on science? Did Darwin inherit better intelligence genes than Watson? This explanation is not very convincing (and would not have been to Darwin). Watson might well have excelled Darwin in academic performance at the university. The three factors which seem most significant were personality, education, and experience.

Personality, Education and Experience

Watson

Watson's difficult personality, on which even *he* commented, is a matter of record. His life experiences and personality traits seem to indicate that he was afflicted with a lifelong personality disorder. It developed within a home environment that was favorable from a material standpoint, but with unlucky circumstances and parents who lacked sophistication in raising their oldest son. Hewett's hostility toward his brothers and hatred toward his father were not passing childhood emotions which he outgrew. These emotions lingered in his psyche, ready to be quickly transferred to new foes, such as Forbes, Prideaux and Trimen. When he was an adult, this became – and was widely recognized as – inappropriate behavior, because he overreacted to minor transgressions. However, there is also another side to this predicament. At a time when the term 'neurosis' included personality disorders, a psychiatrist wrote a best-selling book entitled *Be Glad You're Neurotic*.[1] His thesis is that many people afflicted with lifelong personality problems have made substantial contributions to society, and that many others so afflicted can learn to turn their affliction to their own and to society's advantage. In the cases of Watson and Isaac Newton, one could argue that without their affliction neither would likely have made the substantial contributions to science which they did. Conversely, one could argue that some capable scientists, such as John Stevens Henslow, were 'too normal' to make

---

[1] Bisch 1946.

outstanding contributions to science. They adjusted so well to their situation that they had no inner compulsion to achieve greatness. If Watson's frustration and anger toward his father and brothers had been largely diffused by sophisticated parenting, he might have completed his legal education, entered a legal profession, and remained a Tory and an Anglican. As it happened, however, he was diverted by gardening (this being a step toward sophisticated parenting, but not enough), and this interest later matured and expanded into botanical research. Anger toward his father led him to reject the latter's profession, politics, and religion, but not his wealth, which Hewett invested in his own researches and politics.

Watson was attracted to phrenology largely because of his radical discontent with society (expanded from his discontent with his father and brothers), but he also hoped to gain insights into his own psyche. At Edinburgh he found a favorable environment in which to develop his talents. George Combe served as a mentor. Yet Watson's project of using phrenology to reform society failed because phrenology never gained respectability as a valid science. What he gained for his ego was a belief that he had inherited too large an argumentative faculty. Reason could moderate the expression of this faculty but could not weaken the innate impulses. Even after he gave up on phrenology as a science, he retained it as an ideology that justified his hostile pronouncements on various alleged deficiencies of others. It is a common assumption that Huxley and Darwin's agnosticism represented the farthest extent of unorthodoxy among prominent Victorian scientists; that assumption is challenged by Watson's radical analysis of religious belief.

Rejection of the father was not as thorough as Hewett imagined. Holland and Hewett both undertook managerial roles in society, though their contexts differed. Holland played his role in local governance, and his job involved verbal and legal coercion of people into conformity with the prevailing norms of late Georgian provincial society. Hewett played similar managerial roles within intellectual communities – first phrenological, later botanical – where his self-appointed 'job' involved coercion of people into conformity with his own scientific norms. For both father and son the results were a mix of successes and failures.

Watson's science education was reasonably good for the time. He gained some ability to reason scientifically at the University of Edinburgh, which supplemented the training in argumentation and legal thinking he had acquired from his father and from his apprenticeship. He was sophisticated enough to criticize uncritical thinking among phrenologists. Graham's botany instruction included field trips and the opportunity for Watson to undertake his own studies. In doing so, he discovered the science of plant geography. Although Humboldt was a great earth scientist and plant geographer, his own scientific methodology lacked sophistication. Watson was unable to transcend Humboldt's deficiencies. He read Augustine de Candolle, Lyell, and Herschel without gaining all the important insights which Darwin absorbed from their works. The result was that Watson's innovations fell short of what was needed for the major breakthroughs which he knew were required in biology.

Although Watson's personality disorder may have led him into botany, it also limited his effectiveness as a botanist. In Newton's case, the negative aspects of personality seem less debilitating than in Watson's. Newton's quarrels with Hooke and Leibniz were unproductive; his intellect had better uses. Watson's potential seems even more limited by his affliction. With a less combative personality, he could have maintained a closer relationship with colleagues in science and would not have been marginalized to the extent that he was. Darwin respected his work and welcomed him into his own circle of colleagues, but Watson's personality inhibited him from taking advantage of this opportunity. Watson also failed to become active in The Linnean Society and in the British Association for the Advancement of Science, and in both cases one suspects personality limitations rather than botanical priorities held him back. He was only interested in organizations he could dominate, such as the Botanical Society of London and its exchange club successor.

There is evidence that Watson had a persisting pattern of behavior in which he initially responded favorably to achievements of worthy rivals, but later found reasons for criticism. Examples of this behavior are his responses to Balfour and Babington's *A Catalogue of British Plants*, William Hooker's *The British Flora*, Edward Newman's *History of British Ferns*, and Darwin's theory of evolution by natural selection. This behavioral pattern is also seen in his attitude toward phrenology. His enthusiasm of the 1820s turned to hostility by 1840. Another persisting pattern was a pretense that he was objective, rational and fair concerning the actions and writings of rivals. To lend credibility to this game, he conceded that he was not infallible either, though his performance was always more virtuous than that of his antagonist. This thought pattern probably developed when he was a child defending abusive behavior toward his brothers to a legal-minded father.

These habits of mind help explain why he had a limited capacity to maintain relationships. His most successful and persisting relationships were with George Combe and Grace Eastmond. Watson regarded neither as a rival. Combe was first a mentor, who even then respected Watson's scientific expertise. Later, when Watson left phrenology and Combe more slowly backed away from it, their mutual respect and their correspondence continued. We can only speculate about the quality of Watson's relationship with Eastmond, but it persisted until he died, and the grief she expressed in her letter to Hooker seems genuine. It was never a relationship between equals; Watson was always her employer and she his housekeeper. He had been attracted to Marion Cox, an equal, but he had not tried to woo her away from her role as housekeeper for her uncle, Andrew Combe, or to court her after Andrew died. Watson did persist in a casual relationship with his sisters and their families; again, rivalry never emerged to undermine these relationships.

Among botanists, he succeeded in maintaining relationships only with Joseph Hooker, John G. Baker and Alphonse de Candolle. If there were other botanists with whom his relationship did not deteriorate, it was because their relationships were brief, casual, or where Watson's superiority was acknowledged. William Hooker was a long-time mentor whom Watson eventually rejected because he was not an intellectual, he did not use his government position to advance the botany of the British Isles (as opposed to the world), and he did not keep abreast of advances in British botany well enough to justify (in Watson's opinion) publishing new editions of *The British Flora*. Because Joseph Hooker and John Baker were more than a decade younger than Watson, he did not see them as rivals. In addition, both had particular reasons for not giving up on a relationship with Watson.

For many people, the inheritance of a livelihood leads to indolence or even decline. For neither Watson nor Darwin was this true. Both were diligent researchers and publishers and were respected citizens who contributed to what they saw as good causes. Watson wanted to obtain a good academic appointment in botany, but on two occasions in which he sought a chair, he withdrew his name before decisions were made, because he feared that he would be humiliated by being passed over. It seems unlikely that he would have been a successful professor. He would have been thorough in preparing lectures and conducting field trips, but it is likely that some students or administrators would have aroused his wrath, and that his counter-productive behavior would have followed. He was fortunate in having the independent means that enabled him to avoid being either dismissed from a position or ostracized by colleagues.

Watson's range of science-related experiences seems limited in comparison to Darwin's world travels. In addition to his Azores trip, he traveled widely in Britain, both in the countryside for plants and in cities for phrenological and botanical meetings. He benefited from these experiences, but had no transforming experience comparable to Darwin's voyage of the *Beagle*.

Although Watson may never have consciously concluded that he was a happy man, he had the means to create the life he wanted, and did so. Two misfortunes that befell him in childhood were the stress that distorted his personality, and the cricket accident that permanently damaged one knee. The frozen knee serves as a symbol of his crippled personality. His fall in 1880 led to a painful decline until death. His other tribulations were largely self-inflicted. However, none of them prevented him from carrying out his researches and publications. His failures as phrenological editor and organizer owed about as much to his personality limitations as to the weaknesses of phrenology as a science.

Darwin

Darwin's childhood emotional environment seems better than Watson's. He was a younger son in a liberal family that did not practice primogeniture, and he maintained a close friendship with his older brother, Erasmus, throughout his life. Charles did not turn to nature hobbies as a diversion from family conflict, but simply for amusement. However, his hobbies of beetle collecting and shooting snipe may have been successful sublimations of sexual urges during his teens and early twenties at a time when sexual involvement with 'nice' girls was not available. His enthusiasm for hobbies that could change into biological research was strong enough to make him a successful scientist. His loss of religious belief was a gradual transition that owed much to his increasing involvement in science and to his intellectual sense of security. His high regard for his father and brother, neither of whom was religious, made this an easy step.

There were formal and informal aspects of his science education that were roughly similar to Watson's, and yet Darwin emerged from it with considerably more sophistication than Watson. It is unclear that at Edinburgh Darwin acquired a better understanding than Watson, but then Darwin went on to Cambridge and gained a significant edge. At the emotional level, Combe for Watson and Henslow for Darwin played similar mentoring roles, but at the substantive level Henslow was a competent scientist while Combe was a competent lawyer but no scientist. Darwin also read and admired Humboldt, but Humboldt's methodological influence on him was much less than the combined influence of Whewell, Herschel, Lyell and Augustin de Candolle.[2]

Darwin's five years of travel and researches during the voyage of the *Beagle* was an adventure that he hoped would help him find his way in the world. Because he had a well-prepared mind and conducted comprehensive researches with an open mind, that endeavor succeeded beyond anyone's expectations.

Darwin's vast and fruitful correspondence and his positive interactions with others make it difficult to demonstrate that he had a personality disorder, though his retiring behavior and invalidism after his marriage might provide evidence that he suffered from some kind of emotional affliction. However, the alternative possibilities that he suffered from allergies and from inept medical treatments weaken any argument that his poor health in later years was psychosomatic. Intellectually, he was far more flexible than was Watson. Darwin's extensive family life and colleagueship in science make Watson's social life pale in comparison, though Watson had as much interaction with family and colleagues as he wanted.

---

[2] Ruse 1979, especially chaps. 3 and 7. See also Ghiselin 1969.

Difficult Personalities Within Science

Two types of conflict are inherent in science as a social activity: priority claims and interpretations of evidence. Watson participated in both these types of conflict. An occasional dispute of either type is accepted as normal. However, when a scientist frequently engages in such conflicts, he or she is viewed as a contentious personality. How common is this in science? Watson was the only conspicuously contentious personality within the British botanical community of his day, which was a tolerable level of turbulence for that community. A strain on a scientific community occurs if there is a prolonged conflict between leading scientists in a field or if a conflict becomes highly politicized. The latter situation may occur if conflicting scientists are from different countries and their conflict becomes nationalistic. In Watson's case, no contentious protagonist emerged to prolong any of his conflicts, and so his conflicts were soon forgotten, though everyone remembered that he was contentious.

Scientific Achievements

Although Watson was five years older than Darwin, they both matured within the same intellectual milieu, which included phrenology, evolutionism, Humboldt's environmentalism, Lyell's uniformitarianism, British naval exploration, the growth of scientific institutions, and the professionalization of science.

Watson's false start in phrenology may seem regrettable, but it shows his potential as a theoretical scientist much better than his botanical work does. However, Watson's phrenological speculations were not restrained by reliable scientific evidence, whereas his botanical thoughts were. In botany, he had theoretical ideas which guided his collection of data, but he was too impressed by the limited quantity of that data to use it to test hypotheses and develop theories. As a phrenologist, he was too undisciplined, and as a botanist, too cautious. One suspects that he was also too inarticulate in his own mind about constructing sound hypotheses and theories, and seeing how to test them. We glimpse someone who might have maintained an interaction on an intellectual level with Darwin in the way that Hooker, Huxley and some others did. Such interactions might have helped him learn to hypothesize in botany as actively as he had in phrenology. Instead, his collecting and ordering of data can be compared with that of Tycho Brahe, although unlike Brahe, he lived to analyze his own data. His study of the ratio between areas and the diversity of species was particularly innovative, though a meaningful ecological context for his findings only developed about a century later. His own conclusions were mostly not very impressive; he needed Darwin just as Brahe needed Kepler.

But if Brahe seems unfulfilled without Kepler, the same is less true of Watson in relation to Darwin. Watson's fixation with British botany was stronger than his interest in either evolution or phytogeography. Had he been better at relating to his colleagues, they might have appreciated him more fully, in the way that only de Candolle, Darwin and Baker did while he lived. In the long run, British botanists caught up with his thinking and built upon his diverse achievements.

Greatness in the history of science is estimated mainly on the basis of scientific achievement and influence, with some consideration given to character and behavior. The very highest marks go to scientists like Darwin and Pasteur, who seem outstanding in all these respects. Newton achieved greatness despite a difficult personality and lapses in behavior. Watson was at a lower rank and might fairly be compared with his contemporaries, William and Joseph Hooker. All three were consistently diligent in their efforts to advance botany. The Hookers had personalities that seem normal and reliable in comparison to Watson's, and they were rewarded with important government employment which carried significant scientific influence. Their contributions were both scientific and administrative. Watson had some interest in scientific institutions and contributed to the operation of scientific societies and periodicals, but his main impact was in his writings. Like the Hookers, his name is not associated with some distinctive discovery, such as Darwin's theory of evolution by natural selection or Mendel's laws of heredity. The botanical findings of all three of these botanists were incorporated into the fabric of British and foreign botany and then became indistinct, because the details of what they contributed were too vast to be remembered. But whereas institutional prominence and gentle personalities ensured that the Hookers would be well-remembered, the lack of institutional connection and a difficult personality ensured that Watson would be weakly remembered.

The rediscovery of Mendel's achievement around 1900 rescued him from oblivion. Watson never sank into oblivion, but neither were his achievements sufficiently recognized when the Botanical Society of the British Isles named its journal *Watsonia* and when David Elliston Allen highlighted Watson's contributions within the pages of his history of that society. This colorful figure deserves a more conspicuous place in the history of Victorian science.

Bibliography

Abbreviations

AS	*Annals of Science*
DSB	*Dictionary of Scientific Biography* (Gillispie 1970–80)
CBM	*Companion to the Botanical Magazine*
DNB	*Dictionary of National Biography*
ENPJ	*Edinburgh New Philosophical Journal*
J. Bot.	*Journal of Botany, British and Foreign*
JHB	*Journal of the History of Biology*
L	London
LA	Los Angeles
LJB	*London Journal of Botany*
MNH	*Magazine of Natural History*
MZ&B	*Magazine of Zoology & Botany*
NYC	New York City
P	*Phytologist*
PJ	*Phrenological Journal*
TN	*The Naturalist*

List of Publications by H.C. Watson

Watson, Hewett C. 1829. 'Case of Deficiency of Speech in Concomitance with Great Deficiency in the Organ of Tune', *PJ* 6:103–04.

———. 1830a. 'Evidence Towards Ascertaining the Real Function of Comparison', *PJ* 6:383–97.

———. 1830b. 'Inquiry into the Function of Wit', *PJ* 6:451–69.

[——— H.C.W.] 1830c. 'A Depot for the Exchange of Natural History Articles', *MNH* 3:185.

———. 1831. 'Remarks on the Peculiarities of Memory', *PJ* 7:212–24.

———. 1831–32. 'Account of Experiments on the Weight and Relative Proportions of the Brain, Cerebellum, and Tuber Annulare in Man and Animals, under the Various Circumstances of Age, Sex, Country, &c'. *PJ* 7:434–44.

———. 1832a. *Outlines of the Geographical Distribution of British Plants; belonging to the Division of Vasculares or Cotyledones.* Edinburgh: Neill, xvi + 334 pp.

_____. 1832b. 'Observations' [attached to Graham 1832]. *ENPJ* 13:357–61.

_____. 1833a. 'Observations Made during the Summer of 1832, on the Temperature and Vegetation of the Scottish Highland Mountains, in connexion with Their Height above the Sea', *ENPJ* 14:317–23.

_____. 1833b. 'Observations on the Affinities between Plants and subjacent Rocks', *MNH* 6:424–7.

_____. 1833c. 'On the Relation between Cerebral Development and the Tendency of Particular Pursuits; and on the Heads of Botanists', *PJ* 8:97–108.

_____. 1834a. 'On the Altitude of the Habitats of Plants in Cumberland, with Localities of the Rarer Mountain Species', *MNH* 7:20–4.

_____. 1834b. 'Data Towards Determining the Decrease of Temperature in connexion with the Elevation above the Sea Level, in Britain', *MNH* 7:443–8.

_____. 1835a. *Remarks on the Geographical Distribution of British Plants; Chiefly in Connection with Latitude, Elevation, and Climate*. London: Longman et al., xvi + 288 pp.

_____. 1835b. 'Comparison between the Upper, or Terminal Lines of Trees and Shrubs in Britain, and their Geographical Extension towards the Arctic Regions', *CBM* 1:86–9.

_____. 1835c. 'Numerical Proportions of the Natural Orders of British Plants at Different Elevations', *CBM* 1:196–197.

_____. 1835d. 'Remarks on the Botany of Britain, as Illustrated in Murray's Encyclopaedia of Geography', *CBM* 1:228–34.

_____. 1835e. 'Comments on Mr Hancock's "Letter on the Functions of the Organs of Comparison and Wit"', *PJ* 9:494–8.

_____. 1835–37. *The New Botanist's Guide to the Localities of the Rarer Plants of Britain; on the Plan of Turner and Dillwyn's Botanist's Guide*. Volume 1: *England and Wales*; Volume 2: *Scotland and Adjacent Isles*. London: Longman, xxiv + 674 pp.

_____. 1836a. *Statistics of Phrenology: Being a Sketch of the Progress and Present State of that Science in the British Islands*. London: Longman, x + 242 pp.

_____. 1836b. *An Examination of Mr Scott's Attack upon Mr Combe's "Constitution of Man"*. London: Longman, 39 pp.

_____. 1836c. 'What is the Use of the Double Brain?' *PJ* 9:608–11.

_____. 1836d. 'Statistics of Phrenology', *PJ* 10:54–7.

_____. 1836e. 'Answer to Mr Hancock's Reply to Mr Watson's Comments on his Letter on the Functions of Organs of Comparison and Wit', *PJ* 10:168–70.

_____. 1836f. 'A Fact on the Connection between Mental Emotions and Certain States of the Body', *PJ* 10:283–5.

_____. 1836g. 'Inquiry into the Function of Wit [continued]', *PJ* 10:368–9.

_____. 1836h. 'Observations on the Construction of Maps for Illustrating the

Distribution of Plants, with Reference to the Communication of Mr Hinds on the same Subject', *MNH* 9:17–21.

———. 1837a. 'Perception, Conception, Imagination, and Memory', *PJ* 10: 497–8.

———. 1837b. 'Excitement of Philoprogenitiveness in a Cat', *PJ* 10:725–7.

[———]. 1837c. 'Introductory Explanations to the New Series', *PJ* 11:1–12.

———. 1837d. 'A peculiar Revival of Memory', *PJ* 11:45–7.

———. 1837e. 'Retrospective Strictures', *PJ* 11:69–71.

———. 1837f. 'Why do Birds Sing?' *PJ* 11:73–5.

———. 1837g. 'Magazine of Zoology and Botany', *PJ* 11:66–8.

———. 1837h. 'Observations on the Construction of a Local Flora', *MZ&B* 1:424–30.

———. 1837i. *Bemerkung über die geographische Vertheilung und Verbreitung der Gewächse Grossbritanniens* ... C.T. Beilschmied, transl. Breslau: Jos. Max, xx + 263 pp. [Pritzel no. 10007]

[———]. 1838a. 'Suggestions to Phrenologists, on the Requisites for the Advance of Phrenology', *PJ* 11:97–104.

[——— 'Editor']. 1838b. 'On the Opinions of Phrenologists touching the Function of the Organ called Wit', *PJ* 11:381–91.

———. 1838c. 'The Lowest Temperature of January, 1838'. *TN* 3:241– 3.

———. 1838d. 'Effect of the Winter of 1838 on Vegetation in the Neighbourhood of Thames Ditton, Surrey', *TN* 3:453–7.

[———]. 1839a. 'Phrenology Supported by Scientific Men', *PJ* 12:35–41.

———. 1839b. 'Correspondence between Dr Bardsley, Senior, of Manchester, and Mr Hewett Watson', *PJ* 12:245–8.

[——— Ostensibly by editor Neville Wood]. 1839c. 'The Naturalist's Literary Portrait Gallery. No. III. – Hewett Cottrell Watson, Esq., F.L.S'. *TN* 4:264–9 + portrait.

[———]. 1840a. 'Our Querulous and Critical Correspondents', *PJ* 13:16–29.

[———]. 1840b. [review of] T. Prideaux, *Strictures*, *PJ* 13:264–73.

[———]. 1840c. 'Professor Jameson's Illustrations of a convenient Method for Pruning Reviews that contain inconvenient Opinions', *PJ* 13:303–14.

[———]. 1840d. 'The Phrenological Journal', *PJ* 13:386–7.

———. 1841a. 'Description of a Primula, found at Thames Ditton, Surrey, Exhibiting Characters Both of the Primrose and the Cowslip', *P* 1:9–10.

———. 1841b. Talk with specimens at Botanical Society of London, 1 October 1841, on 5 genera, of which 4 have equivocal species, *P* 1:95–6.

[———]. 1841c. Review of Balfour, Babington & Campbell 1841, *P* 1:109.

———. 1841d. Talk with specimens of *Linaria Bauhini, Lolium multiflorum, Bromus commutatus* at Botanical Society of London, 17 December 1841. *P* 1:136.

———. 1841(?)e. 'Notes on the Distribution of British Ferns', *Bot. Soc. Edinburgh Trans.* 1:89–106. Summarized in *P* 1 (1842) 357–64.

_____. 1842a. *A Letter to the Council of King's College, London*. Thames Ditton: Author, 15 pp.

_____. 1842b. 'A Catalogue of Plants of the Grampians, Viewed in Their Relations to Altitude', *LJB* 1:50–72, 241–54.

_____. 1842c. 'Notices of Some Plants, New To the Flora of Britain', *LJB* 1:76–86.

_____. 1842d. 'Objection to the Alphabetical Arrangement of Local Lists of Plants', *P* 1:150–1.

_____. 1842e. 'Seasons of *Crocus nudiflorus*', *P* 1:188–9.

_____. 1842f. 'The Genus *Tilia*', *P* 1:189.

_____. 1842g. 'Note on the Oxlips from Bardfield, &c'. *P* 1:232–3.

_____. 1843a. *The Geographical Distribution of British Plants*. Edn 3, pt. 1. London: Author, iv + 259 pp. [No more published.]

_____. 1843b. 'Die geographische Verbreitung Britischer Pflanzen, sowohl innerhalb als ausserhalb Grossbritanniens', *Flora* 26: 641–55, 657–71, 681–8, 771–80 and 786–99.

_____. 1843c. 'Remarks on the Distinction of Species in Nature, and in Books; preliminary to the notice of some variations and transitions of character, observed in the native plants of Britain', *LJB* 3:613–22.

_____. 1843d. Note on the Supposed new British *Cerastium*', *P* 1:586–7.

_____. 1843e. 'On the Proposed Change in the Name of *Equisetum limosum*', *P* 1:587–8.

_____. 1843f. 'Places of Growth of *Equisetum fluviatile* of Smith', *P* 1:588.

_____. 1843g. 'The supposed locality of *Geranium nodosum* near Halifax', *P* 1:588–9.

_____. 1843h. 'Unusual Habitat of *Limosella aquatica*', *P* 1:678–9.

_____. 1843i. 'Note on the *Cerastium latifolium* of the *Linnean Herbarium*', *P* 1:717–18.

_____. 1843j. 'The Least Troublesome Method of Drying Plants for the Herbarium', *P* 1:770–1.

_____. 1843k. 'Surrey Localities for *Linaria spartea* and *Senebiera didyma*', *P* 1:775.

_____. 1843l. 'Leaf-buds produced from Roots', *P* 1:776–7.

_____. 1843m. 'Notes on the *Hieracium nigrescens* (Willd) of Babington's *Manual*, and Mr Gibson's *Hieracium hypochoeroides*', *P* 1:801–5.

_____. 1843n. 'A Few Words on the Habitats of *Equisetum Telmateia*', *P* 1:807–8.

_____. 1843–47. 'Notes of a Botanical Tour in the Western Azores', *LJB* 2:1–9, 125–31, 394–408; 3:582–617; 6:380–97.

[_____]. 1844a. *The London Catalogue of British Plants*. Edited for the Botanical Society of London, with assistance of the Secretary, Mr Dennes. London: W. Pamplin. Edn 7, 1874, 1877 and 1881. [Freeman 1980, no. 3917.]

_____. 1844b. 'Notes on the Specific Characters and Varieties of Some British Plants', *LJB* 3:63–81.

_____. 1844c. 'On the Varieties of *Betula alba* of Linnaeus, described as distinct Species by some Authors', *P* 1:821–3.

_____. 1844d. 'Note on Mr Gibson's *Hieracium hypochaeroides*', *P* 1:841.

[_____]. 1844e. 'Notice of *The London Catalogue of British Plants*', *P* 1:932–3.

[_____]. 1844f. 'Analytical Notice, with Illustrations, of a *History of British Ferns and Allied Genera*. By Edward Newman, F.L.S., F.Z.S., &c. London: Van Voorst, 1844', *P* 1:945–960.

_____. 1844g. 'Note on the Bardfield and Claygate Oxlips', *P* 1:1001–2. [W's correction: *P* 2 (1846) 527–8.]

[_____ signed G.E. Dennes]. 1844h. 'Explanations of the *London Catalogue of British Plants*', *P* 1:1014–17, 1098–1100.

_____. 1844i. 'The Nomenclature of Ferns in the *British Flora* and the *London Catalogue of British Plants*', *P* 1:1017–18.

_____. 1844j. Specimens of *Carex elongata* L. and *Bromus commutatus* Schrad. which he collected and indicated locales were exhibited at meeting of Botanical Society of London, 5 July 1844, *P* 1:1062–4.

_____. 1844k. 'Note on the British Species of *Oenanthe*', *P* 1:1083–4.

_____. 1844l. 'Facts about the Nomenclature of Plants in the London and Edinburgh Catalogues', *P* 1:1128–30.

_____. 1844m. 'Note on the Spotted Hieracia', *P* 1:1138–40.

_____. 1845a. 'Botanical Geography of Britain', *LJB* 4:199–208.

_____. 1845b. 'Some Account of the *Œnanthe pimpinelloides*, and *peucedanifolia* of English Authors', *P* 2:11–15.

_____. 1845c. 'Notes on Some British Specimens Distributed by the Botanical Society of London, in 1844–45', *P* 2:43–7.

_____. 1845d. 'On the *Cerastium latifolium* (L.) var *Edmondstonii* (Lond. Cat.); and on the Seeds of *Cerastium latifolium* and *C. alpinum*', *P* 2:93–4.

_____. 1845e. 'Synonyms of *Œnanthe peucedanifolia* of Smith', *P* 2:94–5.

_____. 1845f. 'On the Theory of "Progressive Development", Applied in Explanation of the Origin and Transmutation of Species', *P* 2:108–13, 140–7, 161–8 and 225–8.

_____. 1845g. 'The True Signification of the Term "recurvus"', *P* 2: 170.

_____. 1845h. 'Report of an Experiment which Bears upon the Specific Identity of the Cowslip and Primrose', *P* 2:217–19.

_____. 1845i. 'Observations on Mr. Marshall's Experiments with the Seeds of the Cowslip and Oxlip', *P* 2:313–14.

_____. 1845j. 'Some Words on "Species-making"', *P* 2:314–16.

_____. 1845k. 'On the *Polygonum mite of Schrank*, and Allied Species', *P* 2:332–6.

[_____ signed 'C'.]. 1845–46. 'Notice of the *Annals and Magazine of Natural History*, Nos 100–111', *P* 2:223–4, 478–88.

——, comp. 1846a. *Testimonials in Favour of H.C. Watson, as a Candidate for a Chair of Botany in Ireland*. London: A. Spottiswoode, 51 pp.

——. 1846b. 'Remarks on the Usefulness of a Periodical Devoted to British Botany; Suggested by the *Transactions of the Botanical Society of Edinburgh*', *P* 2:379–83.

——. 1846c. 'Corrections of Various Errors in Mr. Lees' Paper on the *Œnanthe pimpinelloides, Lachenalii and silaifolia*', *P* 2: 390–9.

——. 1846d. 'Correction of Some Errors in the Papers on the Species of *Œnanthe*, in the *Phytologist* for January, 1846 [1845]', *P* 2:405–7.

[—— signed 'C'.] 1846e. 'Notice of A *Flora of Tunbridge Wells* ... By Edward Jenner, A.L.S.', *P* 2:424.

[—— signed 'C'.] 1846f. 'Notice of a *Flora of... Shetland Isles, With Remarks on Their Topography, Geology and Climate*. By Thomas Edmonston ...'. *P* 2:438–40.

——. 1846g. 'Notes on Some British Specimens Distributed by the Botanical Society of London, in 1846', *P* 2:441–4.

[—— signed 'C'.] 1846h. 'Notice of *Flora Azorica*; Founded upon the Collections and Notes of the Two Hochstetters. By Mauritius Seubert, M.D., &c. Bonn, 1844', *P* 2:461–4.

[—— signed 'C'.] 1846i. 'Notice of the London Journal of Botany, Nos. 44, dated August, 1845, to No. 50, dated February 1846', *P* 2:464–7, 477–8.

——. 1846j. 'Note on the English Localities for *Cerastium alpinum*', *P* 2:495.

——. 1846k. 'Notes on the *Ranunculus Leormandi* of Schultz', *P* 2: 497–498.

——. 1846l. 'Correction of a Mistake in Mr. Lawson's 'Stray Thoughts', in the January No. of the *Phytologist*', *P* 2:508.

[—— signed 'C'.] 1846m. 'Notice of *The Vegetable Kingdom; or the Structure, Classification, and Uses of Plants, Illustrated upon the Natural System*. By John Lindley...'. *P* 2:521–6, 594–5.

[—— signed 'Herbarium Committee, Bot. Soc. Lond'.] 1846n. 'Address to Their Fellow Members', *P* 2:538–42.

——. 1846o. 'Notes on the Wild and Cultivated Examples of *Ribes rubrum*', *P* 2:545–8.

——. 1846p. 'Notes on the Lastraea foenesecii as a Species including Both Forms of *Nephrodium foenesecii* (Lowe), and *Aspidium dilatatum*, var. *recurvum* (Bree)', *P* 2:568–71.

——. 1846q. 'Experiment on the Alleged Conversion of the Oat into Rye', *P* 2:605–6.

——. 1846r. 'A Few Words on the First Appearance of Diseased Potatoes in a Garden', *P* 2:613–14.

[—— signed 'C'.] 1846–47. 'Notice of *A Catalogue of the Phaenogamous Plants and Ferns of Great Britain, Arranged According to the Natural Orders; with a Copious List of Synonyms* ... By Henry Ibbotson. Parts I–III', *P* 2:688–95, 878.

_____. 1847a. 'Notes on some British Specimens Distributed by the Botanical Society of London, in 1847', *P* 2:760–8.

_____[signed 'C'.] 1847b. 'Notice of a "Manual of British Botany. By Charles Cardale Babington, M.A., &c". Second Edition', *P* 21:843–51, 871–7.

_____. 1847c. 'Further Experiments Bearing upon the Specific Identity of the Cowslips and Primrose', *P* 2:852–4.

_____. 1847d. Exhibit of a newly-discovered form of an aquatic *Ranunculus* for Britain and various Viola Species at Bot. Soc. London, 7 May, *P* 2:854–5.

[_____ signed 'C'.] 1847e. 'Notice of *The Elements of Botany, Structural and Physiological*, &c. By John Lindley, Ph.D., F.R.S., &c'. *P* 2:866–71.

_____. 1847f. 'Suggestions for Recording the Localities and Distribution of British Plants', *P* 2:893–900.

[_____ signed 'C'.] 1847g. 'Notice of the *London Journal of Botany*, Nos. 65 to 68, Dated May to August, 1847', *P* 2:933–8.

_____. 1847h. 'Notes on the Affinity between *Lysimachia nemorum* (L.) and *Lysimachia azorica* (Hornem.)', *P* 2:975–9.

_____. 1847i. 'On the Credit-worthiness of the Labels Distributed from the Botanical Society of London', *P* 2:1005–15.

_____. 1847j. 'On the *Viola flavicornis* of Smith and others; in Reply to Mr. Forster', *P* 2:1018–22.

[_____ signed 'K'.] 1847k. 'Notice of *Cybele Britannica; or British Plants and Their Geographical Relations*. By Hewett Cottrell Watson', *P* 2:782–800. See also in *LJB* 6:260–2.

_____. 1847–59. *Cybele Britannica; or British Plants, and Their Geographical Relations*. 4 vols. L: Longman. Map in vol 3. [Parts of vols. 1 and 4, selected by FNE, reprinted NYC: Arno Press, 1977.]

_____. 1848a. *Public Opinion; or Safe Revolution through Self Representation*. L: Effingham Wilson, 24 pp.

_____. 1848b. 'On the *Equisetum fluviatile of the London Catalogue of British Plants*', *P* 3:1–4.

[_____ signed 'C'.] 1848c. 'Notice of the *London Journal of Botany*, Nos. 69 to 72, Dated September to December, 1847', *P* 3:4–7.

_____. 1848d. 'Explanations of Some Specimens for Distribution by the Botanical Society of London in 1848', *P* 3:38–49.

[_____ signed 'C'.] 1848e. 'Notice of the *Annals and Magazine of Natural History* for the year 1847, Vols. 19 and 20, or Nos. 123 to 136', *P* 3:49–51.

[_____ signed 'C'.] 1848f. 'Still "Further Remarks" on *Viola flavicornis*, in Reference to Those of Mr. Forster', *P* 3:55–7.

[_____ signed 'C'.] 1848g. 'Notice of *The Flora of Forfarshire*. By William Gardiner. London: Longman & Co. 1848', *P* 3:65–70.

_____. 1848h. 'Is *Gentiana acaulis* Wild in England?' *P* 3:83–4.

_____. 1848i. 'Distribution of *Viola hirta* in Scotland', *P* 3:84–5.

_____. 1848j. 'Reply to Mr. Newman's Quiries on the *Equisetum fluviatile* of the Linnean Herbarium', *P* 3:85–6.

[_____ signed 'C'.] 1848k. 'Notice of the "Tyneside Naturalists Field Club, for the year ending February, 1847", Vol. 1, Part 1. Newcastle, 1848', *P* 3:91–4.

[_____ signed 'C'.] 1848l. 'Notice of the *London Journal of Botany*, Nos. 73 to 75, for January to March, 1848', *P* 3:105–9.

_____. 1848m. 'Reply to Mr. Sidebotham's "Further Remarks on the Second Edition of the *London Catalogue of British Plants*"', *P* 3: 144–5.

_____. 1848n. 'Further Report of Experiments on the Cowslip and Oxlip', *P* 3:146–9.

[_____ signed 'C'.] 1848o. 'Notice of *The Flora of Leicestershire, According to the Natural Orders; Arranged from the London Catalogue of British Plants.* Leicester: John S. Crossly. 1848', *P* 3:157–9.

[_____ signed 'C'.] 1848p. 'Notice of *Contributions towards a Catalogue of Plants Indigenous to the Neighbourhood of Tenby.* London: Longman & Co. 1848', *P* 3:183–4.

[_____ signed 'C'.] 1848q. 'Notice of the *Flora Hertfordiensis ... with the Stations of the Rarer Species.* By the Rev. R.H. Webb, Rector of Essendon; Assisted by the Rev. H. Coleman and by Various Correspondents. Pamplin, London. 1848, Parts I & IV', *P* 3:184–7 & 461–5.

_____. 1848r. 'Characters of *Malva verticillata* and *Malva crispa*', *P* 3:221–2.

_____. 1848s. 'On the Number of Botanical Species to a Square Mile of Ground', *P* 3:267–8.

_____. 1848t. 'Some Account of the Several Alleged Species Included under the Name of *Filago germanica* of Linnaeus', *P* 3:313–18.

_____. 1848u. 'Accidental Introduction of Foreign Plants into Britain', *P* 3:322–3.

_____. 1848v. 'Notice of the *London Journal of Botany*, Nos. 76 to 82, for April to October, 1848', *P* 3:327–30.

_____. 1849a. 'Additions to the Flora of South Wales', *Bot. Gazette* 1:57–9.

[_____ signed 'C'.] 1849b. 'Notice of the *London Journal of Botany*, Nos. 83 and 84, for December, 1848', *P* 3:450–2.

[_____ signed 'O.P.Q'.] 1849c. 'Notice of *A Hand-Book of British Ferns, Intended as a Guide and Companion to Fern Culture.* By Thomas Moore, Curator of the Botanic Garden of the Society of Apothecaries', *P* 3:465–70.

[_____ signed 'K'.] 1849d. 'Notice of *The Elements of Botany.* By M. Adrien de Jussieu, translated by James Hewetson Wilson, F.L.S., F.R.B.S., &c. London: Van Voorst, 1849', *P* 3:471–3.

_____. 1849e. 'Notes on Certain British Plants for Distribution by the Botanical Society of London, in 1849', *P* 3:478–88.

_____. 1849f. 'The Case of the Robertsonian Saxifrages, between Mr. Andrews and Mr. Babington', *P* 3:505–8.

[_____ signed 'C'.] 1849g. 'Reply by the Reviewer 'C'. to Certain Errors of Representation, on the part of Mr. James Blackhouse, Junior, in *Phytologist* iii, pp. 475–6', *P* 3:509–10.

_____. 1849h. 'Reply to Mr. C.C. Babington's Defense, in the Case of the Irish Saxifrages', *P* 3:570.

[_____ signed 'C'.] 1849i. 'Notice of *A Manual of Botany; being an Introduction to the Study of the Structure, Physiology, and Classification of Plants*. By John Hutton Balfour, M.D., &c'. *P* 3:581–6.

_____. 1849j. 'Who Knows Viola *canina*?' *P* 3:635–42.

[_____ signed 'C'.] 1849k. 'Botanical Appointments in the Queen's Colleges, Ireland', *P* 3:642–3.

_____. 1850a. Review: Jules Thurmann, *Essai de phytostatistique* (1849), *Hooker's J. Bot. Kew Gard. Misc.* 2:187–92. See also *P* 3 (1850) 918–21.

[_____ signed 'C'.] 1850b. 'Contents of the *Botanical Gazette* for 1849; a Monthly Journal of Botany, Edited by Arthur Henfrey, F.L.S., &c'. *P* 3:776–83.

_____. 1850c. 'Explanatory Notes on Certain British Plants for Distribution by the Botanical Society of London in 1850', *P* 3:801–11.

[_____ signed 'C'.] 1850d. 'Notice of *A Flora of Leicestershire ...* By Mary Kirby, with Notes by Her Sister, 1850', *P* 3:923–4.

[_____ unsigned]. 1850e. 'Contents of the B*otanical Gazette* No. 17, May, 1850, No. 18, June, 1850', *P* 3:939–40.

_____. 1850f. 'Inquiry for the Celtic or Other Ancient Names of the Doubtfully Native Trees and Shrubs', *P* 3:941–2.

[_____ signed 'C'.] 1850f. 'Notice of *The Tourist's Flora: A Descriptive Catalogue of the Flowering Plants and Ferns of the British Islands, France, Germany, Switzerland, Italy, and the Italian Islands*. By Joseph Woods, F.A.S., &c'. *P* 1042–8.

_____. 1860. *Part First of a Supplement to the Cybele Britannica ...* L: Author, 119 pp. Printed by E. Newman.

_____. 1864a. 'Florula Orcadensis: A List of Plants Reported to Occur in the Orkney Isles. *J. Bot.* 2:11–20.

_____. 1864b. 'Zostera marina in the Orkney Islands', *J. Bot.* 2:54–5.

_____. 1865. 'Calluna vulgaris in Cape Breton, in North America', *Nat. Hist. Rev.*, pp. 149–50.

_____. 1866. 'Corrections in the Shetland Flora', *J. Bot.* 4:348–51.

_____. 1868. 'Chenopodium album, Auct., and Its Varieties', *J. Bot.* 6:289–95.

_____. 1868–70. *A Compendium of the Cybele Britannica; or British Plants in Their Geographical Relations*. L: Longmans, Green, Reader & Dyer, viii + 651 pp. + map. [Title page reads '1870'.]

———. 1869a. '*Chenopodium rubrum*, Linn. Weston Green, Surrey', *J.Bot.* 7:142.
———. 1869b. '*Chenodopium album*, Linn'., *J. Bot.* 7:142–3.
———. 1869c. '*Aira flexuosa* [sic: *uliginosa*], Weihe, in England', *J. Bot.* 7:281–2, [corrected] 337.
———. 1869–70. 'What Is the Thames-side Brassica?' *J. Bot.* 7:346– 50; 8:369–72.
———. 1870. 'Botany of the Azores', pp. 113–288 in Godman 1870.
———. 1871. *A Letter to Mr. John Gilbert Baker, Assistant Editor of the Journal of Botany; Reply to Dr. Trimen's Letter of Defence.* [L:] E. Newman, 8 + 7 pp.
———. 1872. *Supplement to the Compendium of Cybele Britannica; Shewing the Distribution of British Plants Through the Thirty- Eight Sub-Provinces. Being also a Second Supplement to Cybele Britannica.* Thames Ditton: H.C. Watson, iv + 215 pp. + map.
———. 1873–74. *Topographical Botany: Being Local and Personal Records towards Shewing the Distribution of British Plants Traced Through the 112 Counties and Vice-Counties of England, Wales, and Scotland.* Thames Ditton: H.C. Watson, iv + 740 pp. + map.
———. 1881. 'The Authorship of the Third Edition of *English Botany*', *J. Bot.*, n.s. 10:89
———. 1883. *Topographical Botany* ... Edn 2, corrected and enlarged by J.G. Baker & Newbould, W.W. London: Bernard Quaritch, xlvii + 612 pp. + map.

Manuscripts

For guides, see Bridson et al. 1980, and Foster & Sheppard 1989.

Cambridge University Library, Darwin Archive
Darwin, C.R. Letters and manuscripts.

Cambridge University Department of Plant Sciences
Watson, H.C. 1834–61. Letters to C.C. Babington (96).
———. 1847. Letter to Joseph D. Hooker.
———. 1848. Letter to George Lawson.
———. 1859. Letter to John S. Henslow.
———. 1871. Letter to William Mitten.

Cheshire Record Office, Chester
Dawson, Elizabeth. Papers.

Episcopal Registry of Chester
Watson, Holland. 1829. Will.

Royal Botanic Gardens, Kew
_____. 1830–52. Letters to William Hooker (152)
Watson, H.C. 1831. *Gold Medal Prize for Physiological Botany.* Edinburgh, [iii] + 100 + 11 pp.
_____. [1842]. 'Last Will and Testament' (early version), 2 pp. on back of a herbarium sheet of *Eriophorum* in his separately maintained herbarium; photocopy in the Archives.
_____. 1843–63. Letters to George Bentham (7).
_____. 1845–81. Correspondence with Joseph Hooker (80).
_____. 1861? Letters to ? Cooke (2).
_____. 1871–75. Letters to John F. Duthie (6).
_____. 1871–78. Letters to William T. Thiselton-Dyer (4).
_____. 1876. Letters to W.B. Grove (2).
_____. MSS of Topographical Botany and other writings.

British Library, London: Department of Manuscripts
Watson, H.C. 1848. Letter to Robert Peel, the Younger.
Watson, Holland. 1822. Letter to Robert Peel, the Younger.
_____. 1804–26. Letters to Joseph Hunter.

The Linnean Society of London
Watson, H.C. 1832–37. Letters to Nathaniel J. Winch (11).
_____. 1836. Letter to George(?) Don.
_____. 1852. Letter to John E. Gray.
_____. 1861–62. Letters to George Norman (3).

The Natural History Museum, London: Department of Botany
_____. 1861. Letter to G.A. Walker Arnott.
_____. 1865 and 71–5. Letters to John Baker (5 or more).
_____. 186?. Letter to William Carruthers (18 June).
_____. 1874–80. Letters to Frederick A. Lees (2).
_____. 1875–79. Letters to Arthur Bennett (2).
_____. 1880. Letter to James Britten.
_____. 1848. Letter to Thomas Sansom.
_____. 18??. Letter to ? (28 March).
_____. 1832. Letter to R.J. Shuttleworth.

Somerset House
Watson, H.C. 1878. 'Last Will', 2 pp. Proved at London, 20 August 1881, by John Gilbert Baker.

Oxford University: Department of Plant Sciences
Watson, H.C. 1837. Letter to William Pamplin.
_____. 1878–79. Letters to James E. Bagnall (2).
_____. 1878–81. Letters to George C. Druce (5).

University of Newcastle: Robinson Library
Watson, H.C. 1843–73 + n.d. Letters to Walter C. Trevelyan (21).

National Library of Scotland, Edinburgh
Combe, George. 1829–57. Correspondence with H.C. Watson. (includes copies of some of Combe's letters to Watson).

Elgin Museum, Moray, Scotland
Watson, H.C. 1832–62. Letters to George Gordon (5).

Royal Botanic Garden, Edinburgh
Babington, Charles C. 1840–81. Letters to John H. Balfour.
Watson, H.C. 1836–79. Letters to John H. Balfour (31).

Conservatoire et Jardin Botaniques, Geneva, Switzerland
Watson, H.C.? Letter to John T.I. Boswell-Syme.
_____. 1846–74. Letters to Alphonse de Candolle (11).

Biohistorisch Instituut der Rijksuniversiteit, Utrecht, The Netherlands
Watson, H.C. 1875. Letter to John G. Baker.

Hunt Institute for Botanical Documentation, Cornegie Mellon University, Pittsburgh, PA
Watson, H.C. 1878. Letter to James Britten .

Published Works

Abir-am, Pnina G. & Outram, Dorinda, eds. 1987. *Uneasy Careers and Intimate Lives: Women in Science, 1789–1979*. New Brunswick: Rutgers University Press.
Ackerknecht, Erwin H. 1958. 'Contributions of Gall and the Phrenologists to Knowledge of Brain Function', pp. 149–53 in Poynter 1958.

Adie, Alexander. 1819. 'Description of the Patent Sympiesometer or New Air Barometer', *Edinburgh Philos. J.* 1:54–60.
Agassiz, J. Louis. 1840. 'On Glaciers and the Evidence of Their Having Once Existed in Scotland, Ireland, and England', *Geol. Soc. London Proc.* 3:327–32.
Allan, Mea. 1967. *The Hookers of Kew, 1785–1911*. L: Michael Joseph, 273 pp.
———. 1977. *Darwin and His Flowers: The Key to Natural Selection*. L: Faber & Faber, 318 pp.
Allen, David E. 1965. 'H.C. Watson and the Origin of Exchange Clubs', *Proc. Bot. Soc. British Isles* 6:110–112.
———. 1969. *The Victorian Fern Craze: A History of Pteridomania*. L: Hutchinson, xii + 83 pp.
———. 1976. *The Naturalist in Britain: A Social History*. L: Allen Lane, xii + 292 pp.
———. 1980. 'The Women Members of the Botanical Society of London, 1836–1856', *Brit. J. Hist. Sci.* 13:240–54.
———. 1985. 'The Early Professionals in British Natural History', pp. 1–12 in Wheeler & Price 1985.
———. 1986. *The Botanists: A History of the Botanical Society of the British Isles Through a Hundred and Fifty Years*. Winchester: St Paul's Bibliographies, xv + 232 pp.
———. 1999. 'C.C. Babington, Cambridge Botany and the Taxonomy of British Flowering Plants', *Nature in Cambridgeshire* 41:2–11.
——— & Lousley, Dorothy W. 1979. 'Some Letters to Margaret Stovin (1756?–1846), Botanist of Chesterfield', *Naturalist* 104:155–63.
Anon. 1824a. 'Recent Discoveries on the Physiology of the Nervous System', *Edinburgh Med. & Surg. J.* 21:141–59.
———. 1824b. 'On the Combinations in Phrenology, with Specimens of the Combinations of Self-Esteem', *PJ* 1:378–402.
———. 1836. 'Statistics of Phrenology', *PJ*, 10:235–43.
———. 1844. Review of [Watson] 1844a. *LJB* 3:288–95.
——— [Godman?]. 1908. 'Dr. F.D. Godman', *Ibis Jubilee Suppl.* 2:81–92 + photo portrait.
———. 1902. 'Thomas Radcliffe Comber 1837–1902', *J. Bot.* 386–8.
Arnott, G.A. Walker. 1845. 'Corrections of Certain Errors in Dr. Balfour's Communication to the Botanical Society of Glasgow', *P* 2:366–7.
Ashworth, J.H. 1935. 'Charles Darwin as a Student in Edinburgh, 1825–1827', *Proc. Roy. Soc. Edinburgh* 55:97–113.
Athey, Joel. 1988. 'William Hamilton (1788–1856)', p. 349 in Mitchell 1988.
Atran, Scott et al. 1987. *Histoire du concept d'espèce dans les sciences de la vie*. Paris: Foundation Singer-Polignac.

Babington, Anna M., ed. 1897. *Memorials, Journals and Botanical Correspondence of Charles Cardale Babington*. Cambridge: Macmillan & Bowes, xcvi + 476 pp.

Babington, Charles C. 1843. *Manual of British Botany, Containing the Flowering Plants and Ferns Arranged according to the Natural Orders*. L: John Van Voorst, xxiv + 400 pp. Edn 3, 1851. Edn 4, 1856. Edn 5, 1862. Edn 6, 1867. Edn 7, 1874. Edn 8, 1881.

_____. 1844a. 'On Some British Species of Œnanthe', *Ann. Nat. Hist.* 14:96–100; abstract in *P* 1 (1844) 1006; *Edinb. Bot. Soc. Trans.* 2 (1846) 109–112.

_____. 1844b. 'Note on Œnanthe pimpinelloides', *P* 1:1060–1.

_____. 1849a. 'Reply to the Editorial Observations on the Robertsonian Saxifrages, at page 451, &c'. *P* 3:473–5.

_____. 1849b. 'Remarks upon Mr. Watson's Case between Mr. Andrews and Mr. Babington', *P* 3:542–4.

_____. 1850. 'Remarks Resulting from a Perusal of Mr. Watson's *Cybele Britannica*', *Bot. Gaz.* 2:8–10.

_____. 1864. 'Note about the Primula variabilis, Goup.', *J. Bot.* 2:20–1.

Backhouse, James, Jr. 1849. 'A Few Words in Explanation of My "Odd Mistake", as Mentioned by "C".' *P* 3:544–5.

Baker, John G. 1849. 'Occurrence of Carex Persoonii in an Unrecorded Locality in Yorkshire', *P* 3:738–9.

_____. 1855. *The Flowering Plants and Ferns of Great Britain: An Attempt to Classify Them according to Their Geognostic Relations*. L: Cash, 30 pp.

_____. 1875. *Elementary Lessons in Botanical Geography*. L: Lovell Reeve, viii + 110 pp.

_____. 1881. 'In Memory of Hewett Cottrell Watson', *J. Botany* 19:257–65 + photo portrait.

_____. 1883'.In Memory of Hewett Cottrell Watson', reprinted, pp. ix–xxi, in Watson 1883.

_____ & Nowell, John. 1854. *A Supplement to Baines's Flora of Yorkshire*. L: Longman, viii + 188 pp.

Balfour, Isaac B. 1913. 'A Sketch of the Professors of Botany in Edinburgh from 1670 until 1887', pp. 280–301 in Oliver 1913.

Balfour, John H. 1847a. An account of a botanical trip ... with some of his pupils to Clova, Glen Isla, and Braemar', *P* 2:740–1.

_____. 1847b. 'Notes of a botanical trip to the Isle of Wight, in August and September, 1846, with remarks on the geographical distribution of the British Flora', *P* 2:862.

_____. 1848. 'Notes of a Botanical Excursion, with Pupils, to the Mountains of Braemar, Glenisla, and Clova, and to Benlawers, in August 1847', *ENPJ* 45:122–8.

_____. 1849. *A Manual of Botany; being an Introduction to the Study of Structure, Physiology, and Classification of Plants*. Edinburgh: Griffin, xvi + 641 pp. Edn 5, 1859.

_____, Babington, Charles C. & Campbell, W.H., comp. 1841. *A Catalogue of British Plants*. Edn 2. Edinburgh: Maclachlan & Stewart. [Freeman no. 189.]

Ball, John. 1844. 'On Some British Species of the Genus *Œnanthe*', *Ann. Nat. Hist.* 14:4–7; abstract in *P* 1 (1844) 105; *Bot. Soc. Edinb. Trans.* 2 (1846) 105–8.

Barber, Bernard. 1961. 'Resistance by Scientists to Scientific Discovery', *Science* 134:596–602.

Barber, Lynn. 1980. *The Heyday of Natural History*, 1820–1870. L: Jonathan Cape, 320 pp.

Barlow, Nora, ed. 1967. *Darwin and Henslow: The Growth of an Idea; Letters 1831–1860*. L: J. Murray, xii + 251 pp.

Barrett, Paul H. 1973. 'Darwin's "Gigantic Blunder"', *J. Geol. Education* 21:19–28.

Basalla, George. 1972. 'Robert FitzRoy (1805–65)', *DSB* 5:16–18.

Beatty, John. 1985. 'Speaking of Species: Darwin's Strategy', pp. 265–85 in Kohn 1985.

Beesly, Augustus H. 1885. 'Sir George Back (1796–1878)', *DNB*.

Bentham, George. 1858. *Handbook of British Flora*. L: Reeve, xvi + 655 pp. Edn 2, 1865. Edn 3, 1873. Edn 4, 1878.

_____. 1869. 'On Geographical Biology: Anniversary Address', *Linn. Soc. J.* 10:203–206.

_____. 1997. *Autobiography, 1800–1830*. Marion Filipink, ed. Toronto: University Toronto Press, xlviii + 597 pp.

Bickerton, Thomas H. 1936. *A Medical History of Liverpool from the Earliest Days to the Year 1920*. L: J. Murray, xx + 313 pp.

Biermann, Kurt R. 1972. 'Friedrich Wilhelm Heinrich Alexander von Humboldt (1769–1859)', *DSB* 6:549–55.

Bisch, Louis E. 1946. *Be Glad You're Neurotic*. Edn 2, NYC: McGraw-Hill, x + 230 pp.

Blondel, Charles. 1913. *La psycho-physiologie de Gall, ses idées directrices*. Paris: F. Alcan, 166 pp.

Blunt, Wilfrid. 1978. *In for a Penny: A Prospect of Kew Gardens, Their Flora, Fauna, and Falballas*. L: xxi + 218 pp.

Boulger, George S. 1885. 'Berthold Carl Seemann (1825–71)', *DNB*.

_____. 1891. 'Alexander Irvine (1793–1873)', *DNB*.

Bourdier, Franck. 1972. 'Étienne Geoffroy Saint-Hilaire (1772– 1844)', *DSB* 5:355–8.

Bowler, Peter J. 1989. *Evolution: The History of an Idea*. Edn 2. Berkeley and LA: University of California Press, xvi + 432 pp.

_____. 1990. *Charles Darwin: The Man and His Influence*. Cambridge, MA: Basil Blackwell, xii + 250 pp.

Boyd, T. & More, Alexander G. 1858. 'On the Geographical Distribution of Butterflies in Great Britain', *The Zoologist* 16:6018–27. Reprinted with editorial additions by W.F. Kirby in A. More 1898, pp.485–95.

Bradlow, Frank R. 1965. *Baron von Ludwig and the Ludwig's-burg Garden: A Chronicle of the Cape from 1806 to 1848*. Cape Town: Balkema, xii + 124 pp.

[Branch & Son]. 1829. *A Catalogue of a Valuable Library, together with the Book Cases, Iron Safe, and Preserved Birds, of the Late Holland Watson, Esq. To be sold at Auction*. Liverpool: E. Willmer, 16 pp.

Bridson, Gavin D.R., Phillips, Valerie C. & Harvey, A.P., comp. 1980. *Natural History Manuscript Resources in the British Isles*. L: Mansell, xxxv + 473 pp.

Briggs, David & Walters, S.M. 1969. *Plant Variation and Evolution*. NYC: McGraw-Hill, 256 pp.

Britten, James. 1920. 'John Gilbert Baker', *J. Bot.* 58:233–8.

Brock, W.H. 1993. 'Humboldt and the British: A Note on the Character of British Science', *AS* 50:365–72.

Brockway, Lucile H. 1979. *Science and Colonial Expansion: The Role of the British Royal Botanic Gardens*. NYC: Academic Press xvii + 215 pp.

Brooke, John H. 1985. 'The Relations between Darwin's Science and His Religion', pp. 40–75 in Durant 1985.

_____. 1991. *Science and Religion: Some Historical Perspectives*. NYC: Cambridge University Press, x + 422 pp.

Brown, Frank B. 1986. 'The Evolution of Darwin's Theism', *JHB* 19:1–46.

Brown, Robert. 1814. 'General Remarks, Geographical and Systematical, on the Botany of Terra Australis', pp. 533–613 in Flinders 1814. [Reprinted in Brown 1866, vol. 1, pp. 3–89.]

_____. 1818. 'Observations, Systematical and Geographical, on the Herbarium Collected by Professor Christian Smith, in the Vicinity of the Congo, during the Expedition to Explore that River, under the Command of Captain Tuckey, in the Year 1816', pp. 420–85 in Tuckey 1818. [Reprinted in Brown 1866, vol. 1, pp. 98–173.]

_____. 1866. *The Miscellaneous Botanical Works*, ed. John J. Bennett. 2 vols, L: Ray Society.

Browne, E. Janet. 1978. 'The Charles Darwin-Joseph Hooker Correspondence: An Analysis of Manuscript Resources and Their Use in Biography', *J. Soc. Bibliog. Nat. Hist.* 8:351–66.

_____. 1980. 'Darwin's Botanical Arithmetic and the "Principle of Divergence", 1854–1858', *JHB* 13:53–89.

_____. 1981. 'The Making of the *Memoir* of Edward Forbes, F.R.S'., *Arch. Nat. Hist.* 10:205–19.

_____. 1983. *The Secular Ark: Studies in the History of Biogeography*. New Haven, CT: Yale University Press, x + 273 pp.

_____. 1985. 'Darwin and the Expression of the Emotions', pp. 307–26 in Kohn 1985.

_____. 1992. 'A Science of Empire: British Biogeography Before Darwin', *Rev. Hist. Sci.* 45:453–75.

_____. 1995. *Charles Darwin: A Biography. Volume 1: Voyaging.* NYC: Knopf, xv + 605 pp.

Bruhns, Karl, ed. 1872. *Alexander von Humboldt: Eine Wissenschaftliche Biographie.* 3 vols. Leipzig: F.A. Brockhaus.

Budd, Susan. 1977. *Varieties of Unbelief: Atheists and Agnostics in English Society, 1850–1960.* London: Heinemann, vii + 307 pp.

Burkhardt, Richard W., Jr. 1985. 'Darwin on Animal Behavior and Evolution', pp. 327–66 in Kohn 1985.

Burton, Jim. 1986. 'Robert FitzRoy and the Early History of the Meteorological Office', *British J. Hist. Sci.* 19:147–76.

Camerini, Jane. 1997. 'The Power of Biography', *Isis* 88:306–11.

Cameron, George D. 1925. 'Case of Deficient Tune', *PJ* 2:642–4.

Candolle, Alphonse L.P.P. 1847. 'Mémoire sur les causes qui déterminent les limites des espèces du règne végétal du côté du Nord, en Europe, et dans les pays situés d'une manière analogue', *Acad. Sci. Comtes Rendus* 25:895–8.

_____. 1847b. 'On the Causes Which Limit the Distribution of Plants in the North of Europe and Analogous Regions', [transl. A. Henfrey ?], *Bot. Gaz.* 1:113–8.

_____. 1855. *Géographie botanique raisonnée, ou exposition des faits principaux et des lois concernant la distribution géographique des plantes de l'époque actuelle.* 2 vols, Geneva: Kessmann.

_____. 1859. Review of Watson 1847–59, *Archives des sciences physiques et naturelles*, ser. 2, 5:273–8.

_____. 1860. Review of Watson 1847–59, transl. H.C. Watson, pp. 10–18 in Watson 1860.

_____. 1866. 'La vie et les écrits de Sir William Hooker', *Archives des sciences physiques et naturelles*, ser. 2, 25:44–62.

_____. 1869. Review of Moore & More 1866, *Archives des sciences physiques et naturelles*, ser. 2, 36:278–80.

_____. 1880. *Phytographie: ou l'art de décrire les végétaux considérés sous différents points de vue.* Paris: Masson, xxiv + 484 pp.

Candolle, Augustin P. de. 1817. 'Sur la géographie des plantes de France, considérée dans ses rapports avec la hauteur absolue', *Soc. Arcueil, Mém. de Phys.* 3:262–322.

_____. 1820. 'Géographie botanique', *Dictionnaire des sciences naturelles.* Paris: Leverault, 18:359–422. [Reprinted in Egerton 1977.]

Cannon, Susan [=Walter] F. 1977. *Science in Culture: The Early Victorian Period.* NYC: Neale Watson, xii + 290 pp.

Cantor, G.N. 1975a. 'The Edinburgh Phrenology Debate: 1803–1828', *AS* 32:195–218.
_____. 1975b. 'A Critique of Shapin's Social Interpretation of the Edinburgh Phrenology Debate', *AS* 32:245–56.
Castrillon, Alberto. 1992. 'Alexandre de Humboldt et la géographie des plantes', *Rev. Hist. Sci.* 45:419–33.
[Chambers, Robert]. 1844. *Vestiges of the Natural History of Creation*. L: John Churchill, vi + 390 pp. Edn 11, 1860. Edn 12, 1884; facsimile edn 1, NYC: Humanities Press, 1969.
[_____]. 1845. *Explanations: A Sequel to 'Vestiges of the Natural History of Creation'. By the Author of That Work*. L: J. Churchill, 198 pp.
_____. 1994, *Vestiges of the Natural History of Creation and Other Evolutionary Writings*. James A. Secord, ed. Chicago, IL: University of Chicago Press, [xlviii] + vi + 390 + 198 + [199–254] pp.
Chandler, George. 1953. *William Roscoe of Liverpool*. L: B.T. Batsford, xxxvi + 470 pp.
Chitnis, Anand C. 1986. *The Scottish Enlightenment and Early Victorian Society*. L: Croom Helm, v + 201 pp.
Christ, Hermann. 1893. 'Notice biographique sur Alphonse de Candolle', *Bulletin de l'Herbier Boissier* 1:203–34.
Christison, Robert. 1885–86. *The Life of Sir Robert Christison, Bart*. His Sons, eds. 2 vols. Edinburgh: W. Blackwood.
Clark, Ronald W. 1984. *The Survival of Charles Darwin: A Biography of a Man and an Idea*. NYC: Random House, x + 449 pp.
Clarke, Edwin C. & O'Malley, Charles D. 1968. *The Human Brain and Spinal Cord: An Historical Study Illustrated by Writings from Antiquity to the Twentieth Century*. Berkeley: University of California Press.
Coats, Alice M. 1969. *The Quest for Plants: A History of the Horticultural Explorers*. L: Studio Vista, 400 pp.
Coleman, William. 1962. 'Lyell and the "Reality" of Species: 1830–1833', *Isis* 53: 325–38.
Colp, Raph, Jr 1977. *To Be an Invalid: The Illness of Charles Darwin*. Chicago: University of Chicago Press, xiii + 285 pp.
_____. 1980. 'Darwin's "Conspiracy" Against Wallace: Book Review', *New York Academy of Medicine Bulletin* 56:739–48.
_____. 1984. 'The Pre-Beagle Misery of Charles Darwin', *Psychohistory Review* 13:4–15
_____. 1985. 'Notes on Charles Darwin's *Autobiography*', *JHB* 18:357–401.
_____. 1986a. 'Charles Darwin's Dream of His Double Execution', *J. Psychohistory* 13:277–92.
_____. 1986b. 'The Relationship of Charles Darwin to the Ideas of His Grandfather, Dr. Erasmus Darwin', *Biography* 9:1–24.

———. 1987. 'Charles Darwin's "Insufferable Grief"', *Free Associations* 9:6–44.

[Combe, Andrew?]. 1824. 'Flourens on the Nervous System', *PJ* 1:455–63.

———. 1827. 'On the Influence of Organic Size on Energy of Function, particularly as applied to the Organs of the External Senses and Brain', *PJ* 4:161–89.

———. 1834. *The Principles of Physiology Applied to Preservation of Health, and to the Improvement of Physical and Mental Education*. Edn 2. Edinburgh: Adam & Charles Black, viii + 385 pp.

Combe, George. 1819. *Essays on Phrenology, or an Inquiry into the Principles and Utility of the System of Drs Gall and Spurzheim, and into the Objections Made Against It*. Edinburgh: Bell & Bradfute, xxiv + 392 pp.

———. 1828. *The Constitution of Man Considered in Relation to External Objects*. Edinburgh: John Anderson Jr, ix + 319 pp.

———. 1833. *Lectures on Popular Education; Delivered to the Edinburgh Association for Procuring Instruction in Useful and Entertaining Science*. Edinburgh: John Anderson Jr 80 pp.

———. 1836a. *A System of Phrenology*, edn 4, 2 vols. Edinburgh.

———, comp. 1836b. *Testimonials on Behalf of George Combe, as a Candidate for the Chair of Logic in the University of Edinburgh*. Edinburgh: John Anderson Jr, xviii + 79 pp.

[———]. 1845. Review of Chambers 1844. *PJ* 18:69–79.

[———]. 1846. Review of Chambers 1845. *PJ* 19:159–75.

———. 1847a. *The Constitution of Man, Considered in Relation to External Objects*. Edinburgh: Maclachlan & Stewart.

———. 1847b. *On the Relation between Science and Religion*. Edinburgh: Maclachlan & Stewart, i + 45 pp. Edn 4, 1857, xxxi + 280 pp.

———. 1850. *The Life and Correspondence of Andrew Combe, M.D., Fellow of the Royal College of Physicians of Edinburgh*. Edinburgh: Maclachlan & Stewart, ix + 563 pp.

———. 1879. *Education: Its Principles and Practices as Developed by George Combe*. Ed. William B. Jolly. L: Macmillan, lxxvi + 772 pp.

Comber, Thomas. 1873. 'On the World-Distribution of British Plants', *Historic Society of Lancashire & Cheshire Trans.*, ser. 3, 1:237–70.

———. 1874. 'The Dispersion of British Plants', *Historic Society of Lancashire & Cheshire Trans.*, ser. 3, 2:47–74.

———. 1875. 'Geographical Statistics of the Extra-British European Flora', *Historic Society of Lancashire & Cheshire Trans.*, ser, 3. 3:13–32; reprinted *J. Bot*. 1877.

Comrie, John D. 1932. *History of Scottish Medicine to 1860*. Edn 2, 2 vols. L: Wellcome Historical Medical Museum.

Connor, Edward F. & McCoy, Earl D. 1979. 'The Statistics and Biology of the Species–Area Relationship', *Amer. Nat.* 113:791– 833.

Cooter, Roger J. 1976. 'Phrenology: The Provocation of Progress', *History of Science* 14:211–34.

_____. 1984. *The Cultural Meaning of Popular Science: Phrenology and the Organization of Consent in Nineteenth-Century Britain.* NYC: Cambridge University Press, xiv + 418 pp.

Cornell, J.T. 1987. 'God's Magnificent Law: The Bad Influence of Theistic Metaphysics on Darwin's Estimation of Natural Selection', *JHB* 20:381–412.

Courtney, William P. 1901. 'Acton Smee Ayrton (1816–86)', *DNB, Supplement*.

Craigie, David. 1831. 'On William Hamilton's Account of Experiments on the Weight and Relative Proportions of the Brain, Cerebellum, and Tuber Annulare in Man and Animals, under the Various Circumstances of Age, Sex, Country, &c'.. *Edinburgh Med. & Surg. J.* 37:408–18.

Creath, Richard & Maienschein, Jane. 2000. *Biology and Epistemology*. NYC: Cambridge University Press.

Cunningham, Andrew & Jardine, Nicholas, eds. 1990. *Romanticism and the Sciences*. NYC: Cambridge University Press.

Dajoz, Roger. 1984. 'Eléments pour une histoire de l'écologie: La naissance de l'écologie moderne au XIXe siecle', *Histoire et Nature* nos 24–25:5–111 + 7 plts.

Dandy, J.E. 1951. 'The Watsonian Vice-County System', pp.23–30 + plt. in Lousley 1951.

_____. 1969. *Watsonian Vice-Counties of Great Britain*. L: Ray Society, 36 pp. + 2 maps.

Darwin, Charles R. 1839. *Journal of Researches into the Geology and Natural History of the Various Countries Visited by H.M.S. Beagle, under the Command of Captain Fitzroy, R.N., from 1832 to 1836.* L: Colburn, xiv + 615 pp. + 2 maps.

_____. 1846. *Geological Observations on South America, being the Third Part.* L: Smith Elder, vii + 279 pp.

_____. 1857 [1856]. 'On the Action of Sea-water on the Germination of Seeds', *Linn. Soc. London J. (Bot.)* 1:130–40.

_____. 1859. *On the Origin of Species by Means of Natural Selection, or the Preservation of Favoured Races in the Struggle for Life.* L: John Murray, x + 502 pp.

_____. 1862. 'On the Two Forms, or Dimorphic Condition, in the Species of Primula, and on Their Remarkable Sexual Relations', *Linn. Soc. London J. (Bot.)* 6:77–96. [Reprinted in Darwin 1977.]

_____. 1909. *The Foundations of the Origin of Species: Two Essays Written in 1842 and 1844.* Cambridge: Cambridge University Press, xxx + 263 pp. Reprinted NYC: Kraus, 1969.

———. 1959a. *The Autobiography, With Original Omissions Restored*. Nora Barlow, ed. NYC: Harcourt, Brace, 253 pp.

———. 1959b. 'Darwin's Journal', ed. Gavin de Beer, *British Museum (Nat. Hist.) Bull., Hist. Ser.* 2, no. 1, 21 pp.

———. 1959c. *The Origin of Species ...: A Variorum Text*, ed. Morse Peckham. Philadelphia, PA: University of Pennsylvania Press, 816 pp.

———. 1975. *Natural Selection: Being the Second Part of His Big Species Book Written from 1856 to 1858*, ed. R.C. Stauffer. L: Cambridge University Press, xii + 692 pp.

———. 1977. *The Collected Papers*, ed. Paul H. Barrett. 2 vols. Chicago: University of Chicago Press, 312 + 328 pp.

———. 1985–. *The Correspondence*, ed. Frederick Burkhardt, Sydney Smith, Duncan M. Porter, Janet Brown & Marsha Richmond. [1821–63 in 11 vols and continuing], Cambridge University Press.

———. 1987. *Charles Darwin's Notebooks, 1836–1844: Geology, Transmutation of Species, Metaphysical Enquiries*. Barrett, Paul H., Gautrey, Peter J., Herbert, Sandra, Kohn, David & Smith, Sydney, transcribers and eds. Ithaca NY: Cornell University Press. x + 747 pp.

de Candolle: *see* Candolle.

de Giustino, David A. 1975. *Conquest of Mind: Phrenology and Victorian Social Thought*. Totowa, NJ: Rowman & Littlefield.

Delaunay, Paul. 1953. 'Les doctrines médicales au début du XIXe: Louis et la méthode numérique', vol. 2, pp. 3211–330 in Underwood 1953.

Dennes, George E. 1844. *see* Watson 1844h.

Desmond, Adrian. 1984. 'Robert E. Grant: The Social Predicament of a Pre-Darwinian Transmutationist', *JHB* 17:189–223.

———. 1985. 'Richard Owen's Reaction to Transmutation in the 1830s', *Brit. J. Hist. Sci.* 18:25–50.

———. 1987. 'Artisan Resistance and Evolution in Britain, 1819–1848', *Osiris*, ser. 2, 3:77–110.

———. 1989. 'Lamarckism and Democracy: Corporations, Corruption and Comparative Anatomy in the 1830s', pp. 99–130 in Moore 1989.

———. 1997. *Huxley: From Devil's Disciple to Evolution's High Priest*. Reading MA: Addison-Wesley, xxi + 820 pp.

——— & Moore, James. 1991. *Darwin*. NYC: Warner, xxi + 808 pp.

Desmond, Ray. 1970a. 'Charles Cardale Babington (1808–95)', *DSB* 1:358–9.

———. 1970b. 'John Gilbert Baker (1834–1920)', *DSB* 1:412–3.

———. 1970c. 'John Hutton Balfour (1808–84)', *DSB* 1:423.

———. 1977. *Dictionary of British and Irish Botanists and Horticulturists, including Plant Collectors and Botanical Artists*. L: Taylor & Francis, xxvi + 747 pp.

———. 1995. *Kew: The History of the Royal Botanic Gardens*. L: Harvill Press, xvii + 466 pp.
Dettelbach, Michael. 1996. 'Global Physics and Aesthetic Empire: Humboldt's Physical Portrait of the Tropics', pp. 258–92 in Miller & Reill 1996.
Dickie, George. 1843a. 'Notes on the Distribution of the Plants of Aberdeenshire in Relation to Altitude', *LJB* 2:131–5.
———. 1843b. 'Second Paper on the Distribution of Aberdeenshire Plants', *LJB* 2:355–8.
———. 1847. 'Notes of Algae, Observed at Various Altitudes in Aberdeenshire', *LJB* 6:197–206, 376–80.
Dobbs, Betty Jo T. 1994. Reviews of: A.R. Hall, *Isaac Newton* (1992) and R. Westfall, *The Life of Isaac Newton* (1993), *Isis* 85:515–7.
Dony, J.G. 1963. 'The Expectation of Plant Records from Prescribed Areas', *Watsonia* 5:377–85.
Drouin, Jean Marc. 1991. 'Quelques figures de l'insularité: Réflexions sur la biogéographie', pp. 197–216 in Roger & Guéry 1991.
———. 1993. *L'écologie et son histoire: Réinventer la nature*. [Paris]: Flammarion, 218 pp.
———. 1994. 'La regle et l'ecart: La philosophie botanique d'Augustin-Pyramus de Candolle', *Ludus Vitalis* 2:69–91.
———. 1995. 'Quantification and Plant-Geography in the 19th Century, 1815–1855', unpublished MS, 10 pp.
Druce, George C. 1932. *The Comital Flora of the British Isles ... Being the Distribution of British (Including a Number of Non-Indigenous) Plants Throughout the 152 Vice-Counties of Great Britain, Ireland, and the Channel Islands, with the Place of Growth, Elevation, World-Distribution ... With an Original Coloured May Showing the Botanical Vice-Counties presented by William James Patey*. Arbroath: Buncle, xxxii + 407 pp.
Du Rietz, G. Einar. 1957. 'Linnaeus as a Phytogeographer', *Vegetatio* 7:161–8.
Durant, John, ed. 1985. *Darwinism and Divinity*. Oxford: Basil Blackwell, 1985.
Edsall, Nicholas C. 1986. *Richard Cobden: Independent Radical*. Cambridge, MA: Harvard University Press, xiv + 465 pp.
Egerton, Frank N. 1968. 'Studies of Animal Populations from Lamarck to Darwin', *JHB* 1:225–59.
———. 1970a. 'Humboldt, Darwin, and Population', *JHB* 3:325–60.
———. 1970b. 'Refutation and Conjecture: Darwin's Response to Sedgwick's Attack on Chambers', *Stud. Hist. Philos. Sci.* 1:179–81.
———. 1971. 'The Concept of Competition in Nature before Darwin', *XIIe Congrès International d'Histoire des Sciences, Actes* 8:41–6.
———. 1972. 'Edward Forbes, Jr (1815–54)', *DSB* 5:66–8.

_____. 1974. 'Charles Darwin's Analysis of Biological Competition', *13th International Congress of the History of Science* 9:71–7.

_____, comp. 1977. *Ecological Phytogeography in the Nineteenth Century*. NYC: Arno Press.

_____. 1979. 'Hewett C. Watson, Great Britain's First Phytogeographer', *Huntia* 3:87–102.

_____. 1995. 'In Quest of a Science: Hewett Watson and Early Victorian Phrenology', *Essays in Arts and Sciences* 24:1–20.

Eiseley, Loren C. 1979. *Darwin and the Mysterious Mr. X: New Light on the Evolutionists*. NYC: E.P. Dutton, xiii + 278 pp.

Eisen, Sydney & Lightman, Bernard V., comps. 1984. *Victorian Science and Religion: A Bibliography on Evolution, Belief, and Unbelief*. Hamden, CT: Archon Books, xix + 696 pp.

Eriksson, Gunnar. 1976. Göran Wahlenberg (1780–1851)', *DSB* 14:116–7.

Eyde, Richard H. 1975. 'The Foliar Theory of the Flower', *Amer. Scientist* 63:430–7.

Farmer, J.B. 1913. 'Robert Brown, 1773–1858', pp. 108–25 in Oliver 1913.

[Fleming, John]. 1854. 'Botanical Geography', *North British Review* 20:501–22.

Fletcher, Harold R. & Brown, William H. 1970. *The Royal Botanic Garden, Edinburgh, 1670–1970*. Edinburgh: Her Majesty's Stationery Office, xvi + 309 pp.

Flinders, Matthew. 1814. *A Voyage to Terra Australis; Undertaken for the purpose of Complelting the Discovery of that vast country ... 1801, 1802 and 1803*. 2 vols. L.

Flourens, Pierre. 1824. *Researches expérimentales sur les propriétés et les fonctions du système nerveux dans les animaux vertébrés*. Paris: Crevot. Edn 2, 1842.

Forbes, Edward. 1845. 'On the Distribution of Endemic Plants, More especially Those of the British Islands, Considered with Regard to Geological Changes', *Brit. Assoc. Advan. Sci. Report* 15:67–68; *Ann. Mag. Nat. Hist.* 16:126–7; *Athenaeum* no. 923: 678; *Gard. Chron.* no. 28:473–4; *Lit. Gaz.* no. 1484:414–15; *P* 2:272–3.

_____. 1846a. 'On the Connexion between the Distribution of the Existing Fauna and Flora of the British Isles and the Geological Changes which Have Affected Their Area', *Memoirs of Geol. Soc. of England & Wales* 1:336–432.

_____. 1846b. 'Descriptions of Secondary Fossil Shells from South America', pp. 265–8 + plt. 5 in Darwin 1846.

_____. 1852. 'Plants and Botanists', *Westminster Rev.* n.s. 2:385–98; Amer. ed., pp. ?–212–?

Foster, Janet & Sheppard, Julia, eds. 1989. *British Archives: A Guide to Archive Records in the United Kingdom*. edn 2, NYC: Stockton P., lviii + 834 pp.

Foster, W.D. 1972. Review of Winslow 1971, *Isis* 63:591–2.

Fraser, Robert S. 1988. 'Lord John Russell (1792–1878)', pp. 687–8 in Mitchell 1988.

Freeman, Richard B. 1865. *The Works of Charles Darwin: An Annotated Bibliographical Handlist*. Edn 2. Hamden CT: Archon Books, 235 pp.

_____. 1978. *Charles Darwin: A Companion*. Folkestone: W. Dawson & Sons, 309 pp.
_____. 1980. *British Natural History Books, 1495–1900: A Handlist*. L: W. Dawson & Sons, 437 pp.
Gage, Andrew T. & Stearn, William T. 1988. *A Bicentenary History of the Linnean Society of London*. L: Academic Press, ix + 242 pp.
Gajewski, W. 1959. 'Evolution in the Genus Geum', *Evolution* 13:378–88.
Gall, Franz J. & Spurzheim, Johann C. 1810–19. *Anatomie et physiologie du système nerveux en général, et du cerveau en particuliere, avec des observations sur la possibilité de reconnoître plusieurs dispositions intellectuelles et morales de l'homme et des animaux, par la configuration de leurs têtes*. 4 vols. + atlas. Paris: F. Schoell.
_____ & _____. 1822–25. *Sur les fonctions du cerveau et sur celles de chacune de ses parties*. 6 vols, Paris. [Edn 2 of 1810–19.]
_____ & _____. 1835. *On the Functions of the Brain and of Each of Its Parts: With Observations on the Possibility of Determining the Instincts, Propensities, and Talents, or the Moral and Intellectual Dispositions of Men and Animals, by the Configuration of the Brain and Head*. Transl. W. Lewis Jr. 6 vols, Boston, MA: Marsh, Capen & Lyon.
Gardiner, Brian G. 1993. 'Edward Forbes, Richard Owen and the Red Lions', *Archives of Nat. Hist.* 20:349–72.
Gash, Norman. 1979. *Aristocracy and People: Britain, 1815–1865*. Cambridge MA: Harvard University Press, 375 pp.
Gaukroger, Stephen. 1995. *Descartes: An Intellectual Biography*. Oxford: Clarendon P., xx + 499 pp.
Geison, Gerald L. 1969. 'Darwin and Heredity: The Evolution of His Hypothesis of Pangenesis', *J. Hist. Med. Allied Sci.* 24: 375–411.
_____. 1972. 'Arthur Henfrey (1819–59)', *DSB* 6:265–7.
George, Wilma. 1964. *Biologist Philosopher: A Study of the Life and Writings of Alfred Russel Wallace*. NYC: Abelard-Schuman, x + 320 pp.
Ghiselin, Michael T. 1969. *The Triumph of the Darwinian Method*. Berkeley, CA & LA, CA: University of California Press, vii + 287 pp.
_____. 1973a. 'Mr. Darwin's Critics, Old and New', *JHB* 6:155–65.
_____. 1973b. 'Darwin and Evolutionary Psychology', *Science* 179: 964–8.
_____. 1976. 'Two Darwins: History versus Criticism', *JHB* 9:121–32.
Gibbon, Charles. 1878. *The Life of George Combe, Author of 'The Constitution of Man'*. 2 vols, L: Macmillan.
Gilbert, Alan D. 1980. *The Making of Post-Christian Britain: A History of the Secularization of Modern Society*. L: Longman, xv + 173 pp.
Gillespie, Neal C. 1979. *Charles Darwin and the Problem of Creation*. Chicago: University of Chicago Press, xiii + 201 pp.

Gillispie, Charles C. 1951. *Genesis and Geology: A Study in the Relations of Scientific Thought, Natural Theology, and Social Opinion in Great Britain, 1790–1850*. Cambridge, MA: Harvard University Press, xv + 315 pp.

_____, ed. 1970–80. *Dictionary of Scientific Biography*. 16 vols, NYC: Charles Scribner's Sons.

Gilmour, John S. 1963. 'The Changing Pattern', pp. 9–13 in Wanstall 1963.

Glen, Robert. 1984. *Urban Workers in the Industrial Revolution*. NYC: St Martin's Press, vii + 348 pp.

_____. 1988. 'Manchester', pp. 474–5 in Mitchell 1988.

Godman, Frederick D. 1870. *Natural History of the Azores, or Western Islands*. L: John Van Voorst, viii + 358 pp.

Gordon, Alexander. 1891. 'William Ballantyne Hodgson (1815– 1880)', *DNB*.

[Gordon, John] 1815. Review of: Gall & Spurzheim, *Anatomie et physiologie* (2 vols, 1810–12) and Spurzheim, *Physiognomial System* (edn 2, 1815), *Edinburgh Review* 25:227–68.

_____. 1817. *Observations on the Structure of the Brain, comprising an Estimate of the Claims of Drs Gall and Spurzheim to Discovery in the Anatomy of that Organ*. Edinburgh: Blackwood.

Gorham, Eville. 1954. 'An Early View of the Relation between Plant Distribution and Environmental Factors', *Ecology* 35:97– 8.

Graham, Robert. 1832. 'Botanical Excursions into the Highlands of Scotland from Edinburgh this Season', *ENPJ* 13:350–7.

_____. 1833. 'Notice of Botanical Excursions into the Highlands of Scotland from Edinburgh this Season, 1833', *ENPJ* 15:358–61.

_____. 1835. 'Plants New to the British Flora, or Rare in Scotland', *ENPJ* 19:346–51.

Grant, A. Cameron. 1965. 'Combe on Phrenology and Free Will: A Note on XIXth-Century Secularism', *J. Hist. Ideas* 26:141–7.

Grant, Alexander. 1884. *The Story of the University of Edinburgh during Its First Three Hundred Years*. 2 vols. L: Longmans, Green.

Gray, J. M. 1893. 'Lawrence Macdonald (1799–1878)', *DNB*.

Greene, Edward Lee. 1983. *Landmarks of Botanical History*. Ed. Frank N. Egerton. 2 vols. Stanford: Stanford University Press.

Greene, John C. 1981. *Science, Ideology, and World View. Essays in the History of Evolutionary Ideas*. Berkeley, CA and LA, CA: University of California Press, x + 202 pp.

Gregory, Michael S., Silvers, Anita & Sutch, Diane, eds. 1978. *Sociobiology and Human Nature*. San Francisco, CA: Jossey-Bass.

Grindon, Leo H. 1844. 'A Few More Words on *The London Catalogue*', *P* 1:1077–9.

_____. 1877. *Manchester Banks and Bankers: Historical, Biographical, and Anecdotal*. Manchester: Palmer & Howe, viii + 333 pp.

Grisebach, August H.R. 1846a. 'Report on the Contributions to Botanical Geography, during the Year 1842', transl. W.B. MacDonald, pp. 55–122 in *Reports and Papers on Botany*. L: Ray Society. [Reprinted in Egerton 1977.]

———. 1846b. 'Report on the Contributions to Botanical Geography, during the Year 1843', transl. George Busk, pp. 123–12 in Grisebach 1846a. [Reprinted in Egerton 1977.]

———. 1849a. 'Report on Geographical Botany for 1844', pp. 317–413 in Arthur Henfrey, ed., *Reports and Papers on Botany*. L: Ray Society. [Reprinted in Egerton 1977.]

———. 1849b. 'Report on the Progress of Geographical and Systematic Botany during the Year 1845', pp. 415–73 in Grisebach 1849a. [Reprinted in Egerton 1977]

———. 1872a. *Die Vegetation der Erde nach ihrer klimatischen Anordnung: Ein Abriss der vergleichenden Geographie der Pflanzen*. 2 vols. Leipzig: Engelmann.

———. 1872b. 'Pflanzengeographie und Botanik', vol. 3, pp. 232–68 in Bruhns 1872. [Not included in English translation.]

Gruber, Howard E. & Barrett, Paul H. 1974. *Darwin on Man*. NYC: Dutton, xxx + 495 pp.

Guimond, Alice A. 1972. 'William Herbert (1778–1847)', *DSB* 6:295–7.

Gunther, A.E. 1974. 'A Note on the Autobiographical Manuscripts of John Edward Gray (1800–1875)', *Soc. Bibliog. Nat. Hist. J.* 7:35–76.

———. 1975. *A Century of Zoology at the British Museum through the Lives of Two Keepers*, 1815–1914. L: Dawsons of Pall Mall, 533 pp.

Guppy, Henry B. 1917. *Plants, Seeds and Currents in the West Indies and the Azores: The Results of Investigations Carried Out in Those Regions between 1906 and 1914*. L: Williams & Norgate, xi + 531 pp.

Gwynn, Denis. 1948. *O'Connell, Davis and the Colleges Bill*. Oxford: Blackwell, iii + 88 pp.

Hall, A. Rupert. 1980. *Philosophers at War: The Quarrel Between Newton and Leibniz*. NYC: Cambridge University Press, xiii + 338 pp.

Hamilton, John A. 1891. 'Joseph Hume (1777–1855)', *DNB*.

Hamilton, William; Spurzheim, J. & Combe, G. [anon.]. 1828. 'Sir William Hamilton, Bart., and Phrenology', *PJ* 5:1–69, 153–8.

Hancock, George. 1835. 'Letter on the Functions of the Organs of Comparison and Wit', *PJ* 9:435–43.

———. 1836. 'Reply to Mr H.C. Watson's Comments on his Letter on the Functions of Comarison and Wit', *PJ* 10:14–19.

Harrington, Anne. 1987. *Medicine, Mind, and the Double Brain*. Princeton, NJ: Princeton University Press, xiii + 336 pp.

Head, Robert. 1887. *Congleton: Past and Present*. Congleton: Robert Head, xx + 290 pp.

Heath, John & Perring, Franklyn. September 1975. 'Biological Recording in Europe', *Endeavour* 34: 103–8.
Henderson, Thomas F. 1885. 'James Abercromby, Lord Dunfermline (1776–1858)', *DNB*.
———. 1887. 'Robert Cox (1810–72)', *DNB*.
Henfrey, Arthur. 1849. 'Discrimination of Species', *P* 3:488–9.
———. 1852. *The Vegetation of Europe, Its Conditions and Causes*. L: John Van Voorst, ii + 387 pp. + map.
Henslow, John S. 1830. 'On the Specific Identity of the Primrose, Oxlip, Cowslip, and Polyanthus', *MNH* 3:406–9.
Hepper, F. Nigel, ed. 1982. *Royal Botanic Gardens Kew: Gardens for Science & Pleasure*. Owings Mills MD: Stemmer House, vii + 195 pp.
Herbert, William. 1822. 'On the Production of Hybrid Vegetables; With the Result of Many Experiments Made in the Investigation of the Subject', *Hort. Soc. London Trans.* 4:15–47.
Herschel, John F.W. 1830. *A Preliminary Discourse on the Study of Natural Philosophy*. L.: Longman, Rees, Orme, Brown & Green, viii + 372 pp.
Hewins, W.A.S. 1893. 'Sir George Stewart Mackenzie (1780–1848)', *DNB*.
Heyck, Thomas W. 1982. *The Transformation of Intellectual Life in Victorian England*. NYC: St Martin's Press, 262 pp.
Hibbert, Christopher. 1978. *The Great Mutiny: India 1857*. NYC: Viking Press, 472 pp.
Hinde, Wendy. 1987. *Richard Cobden: A Victorian Outsider*. New Haven, CT: Yale University Press, xi + 367 pp.
Hodge, M.J.S. 1972. 'The Universal Gestation of Nature: Chambers' *Vestiges* and *Explanations*', *JHB* 5:127–52.
———. 1977. 'The Structure and Strategy of Darwin's "Long Argument"', *JHB* 10:237–46.
———. 1987. 'Darwin, Species and the Theory of Natural Selection', pp. 227–52 in Atran et al. 1987.
Hooker, Joseph D. 1851a. 'Enumeration of the Plants of the Galapagos Islands, with Descriptions of the New Species', *Linnean Soc. Lond. Trans.* 20:163–233. Summary: *P* 3 (1848) 24–8.
———. 1851b. 'On the Vegetation of the Galapagos Archipelago, as Compared with That of Some Other Tropical Islands and of the Continent of America', *Linnean Soc. Lond. Trans.* 20:235–62. Summary: *P* 3 (1848) 24–8.
———. 1862. 'Outlines on the Distribution of Arctic Plants', *Linn. Soc. Lond. Trans.* 23:251–348 + map.
———. 1867. 'On Insular Floras', *J. Bot.* 5:23–31.
———. 1870. *The Student's Flora of the British Islands*. L: Macmillan, xx + 504 pp. Edn 2, 1878. edn 3, 1884.

_____. 1903. *A Sketch of the Life and Labours of Sir William Jackson Hooker, K.H., Late Director of the Royal Gardens of Kew.* Oxford: Clarendon Press, pp. ix–xc.

_____. 1953. *Pioneer Plant Geography: The Phytogeographical Researches*, ed. William B. Turrill. The Hague: M. Nijhoff, xii + 267 pp.

Hooker, William J. 1821. *Flora Scotica; or a Description of Scottish Plants, arranged Both according to the Artificial and Natural Methods.* Edinburgh: Archibald Constable, 296 + 303 pp.

_____. 1830. *The British Flora: Comprising the Phaenogamous, or Flowering Plants, and the Ferns.* L: Longman x + 480 pp.

_____. 1834. 'Geography Considered in Relation to the Distribution of Plants', pp. 227–46 in Murray et al. 1834. [Reprinted in Egerton 1977.]

[_____.] 1835. Review of Watson 1835a. *Companion to Bot. Mag.* 1:195.

_____, comp. 1844. *Icones Plantarum; or Figures with Brief Descriptive Characters and Remarks, of New or Rare Plants, Selected from the Author's Herbarium*, vol. 7. L: H. Baillière.

_____ & Arnott, George A.W. 1850. *The British Flora: Comprising the Phaenogamous or Flowering plants, and the Ferns.* Edn 6, L: Longman, xil + 604 pp.

_____ & _____. 1855. Ed. 7, xliv + 618 pp.

Hoyningen-Huene, Paul. 1993. *Reconstructing Scientific Revolutions.* Chicago: University of Chicago Press, xx + 310 pp.

Hull, David L., comp. & ed. 1973. *Darwin and His Critics: The Reception of Darwin's Theory of Evolution by the Scientific Community.* Cambridge, MA: Harvard University Press, 480 pp.

_____. 1978. 'Scientific Bandwagon or Traveling Medicine Show?'. pp. 136–63 in Gregory, Silvers & Sutch 1978.

_____. 2000. 'Why Did Darwin Fail? The Role of John Stuart Mill', pp. 48–63 in Creath & Maienschein 2000.

Humboldt, F.W.H. Alexander von. 1817a. *De Distributione Geographica Plantarum Secundum Coeli Temperiem et Altitudinem Montium, Prolegomena.* Paris: Libraria Graeco-Latino-Germanica, 247 + 2 + 2 pp.

_____. 1817b. 'Des lignes isothermes et la distribution de la chaleur sur le globe', *Arcueil, Mém. de Phys.* 3:462–602.

_____. 1820–21. 'On Isothermal Lines, and the Distribution of Heat over the Globe', *Edinburgh Philos. J.* 3:1–20, 256–74; 4:23–37, 262–81; 5:28–39. [Reprinted in Egerton 1977.]

_____ & Bonpland, Aimé. 1807 [not 1805]. *Essai sur la géographie des plantes; accompagné d'un tableau physique des régions équinoxi- ales, Fondé sur des mesures exécuté, depuis le dixième degré de latitude boréale jusqu'au dixième degré australe, pendant les années 1799, 1800, 1801, 1802 et 1803.* Paris: Levrault, Schoell, 155 pp. + chart.

_____ & _____. 1814–29. *Personal Narrative of Travels to the Equinoctial Regions of the New Continent, during the Years 1799–1804*. Transl. Helen Maria Williams. 7 vols. L: Longman. [Reprinted, NYC: AMS Press, 1966.]

Hunt, T. Carew. 1845a. 'A Description of the Island of St. Mary (Azores)', *Roy. Geogr. Soc. J.* 15:258–68 + map.

_____. 1845b. 'A Description of the Island of St. Michael (Azores)', *Roy. Geogr. Soc. J.* 15:268–96 + map.

Huxley, Leonard. 1918. *Life and Letters of Sir Joseph Dalton Hooker, O.M., G.C.S.I., Based on Materials Collected and Arranged by Lady Hooker*. 2 vols. L: Macmillan.

Inkster, Ian & Morrell, Jack, eds. 1983. *Metropolis and Province: Science in British Culture, 1780–1850*. L: Hutchinson, 288 pp.

Irvine, William. 1955. *Apes, Angels, and Victorians*. NYC: McGraw-Hill, xv + 399 pp.

Jackson, B. Daydon. 1883. 'Local Catalogues Used in Preparing Watson's "Topographical Botany"', *J.Bot.* 21:343–6, 363–70.

_____. 1906. *George Bentham*. L: Dent, xii + 292 pp.

Jarrett, Derek. 1989. *The Sleep of Reason: Fantasy and Reality from the Victorian Age to the First World War*. NYC: Harper & Row, ii + 233 pp. [First published in England, 1988.]

Kargon, Robert H. 1977. *Science in Victorian Manchester: Enterprise and Expertise*. Baltimore, MD: Johns Hopkins University Press, 288 pp.

Kelham, Brian B. 1973. 'Robert John Kane (1809–90)', *DSB* 7:224.

Kitteringham, Guy. 1982. 'Science in Provincial Society: The Case of Liverpool in the Early 19th Century', *AS* 39:329–48.

Kohn, David, ed. 1985. *The Darwinian Heritage*. Princeton NJ: Princeton University Press, xii + 1138 pp.

_____, Smith, Sydney & Stauffer, R.C. 1982. 'New Light on The Foundations of the Origin of Species: A Reconstruction of the Archival Record', *JHB* 15:419–42.

Kottler, Malcolm J. 1974. 'Alfred Russel Wallace, the Origin of Man, and Spiritualism', *Isis* 65:145–92.

Kuhn, Thomas S. 1970. *The Structure of Scientific Revolutions*. Edn 2, Chicago: University Chicago Press.

Lakatos, Imre & Musgrave, Alan, eds. 1970. *Criticism and the Growth of Knowledge*. NYC: Cambridge University Press.

Lanteri-Laura, Georges. 1970. *Histoire de la phrénologie: L'homme et son cerveau selon F.J. Gall*. Paris: Presses Universités de France.

Large, Ernest C. 1940. *The Advance of the Fungi*. L: Jonathan Cape, 488 pp.

Larson, James. 1986. 'Not without a Plan: Geography and Natural History in the Late Eighteenth Century', *JHB* 19:447–88.

Lawrence, George H.M., Buchheim, A.F.G., Daniels, G.S. & Dolezal, Helmut, eds. 1968. *B-P-H: Botanico-Periodicum-Huntianum*. Pittsburgh': Hunt Botanical Library, 1063 pp.
Layton, David. 1981. 'The Schooling of Science in England, 1854–1939', pp. 188–210 in MacLeod & Collins 1981.
Lecoq, Henri. 1854–58. *Études sur la géographie botanique de l'Europe et en particulier sur la végétation du plateau central de la France*. 9 vols. Paris: H.Baillière.
Lesky, Erna. 1970. 'Structure and Function in Gall', *Bull. Hist. Med.* 44:297–314.
Lewes, George H. 1867. *The History of Philosophy from Thales to Comte*. Ed. 3, 2 vols. L: Longmans, Green.
Lindberg, David C. & Numbers, Ronald L., eds. 1986. *God and Nature: Historical Essays on the Encounter between Christianity and Science*. Berkeley: University California Press, xi + 516 pp.
Lindley, John. 1829. *A Synopsis of the British Flora; Arranged according to the Natural Orders: Containing Vasculares, or Flowering Plants*. L: Longman et al., xii + 360 pp.

———. 1830. *An Introduction to the Natural System of Botany: or, a Systematic View of the Organization, Natural Affinities, and Geographical Distribution of the Whole Vegetable Kingdom*. L.: Longman, xlviii + 376 pp.

[———]. 1859. Review of volume 4 of Watson 1847–59. *Gardners Chronicle*, no. 46:911–12.
Linnaeus, Carl. 1972. *L'Équilibre de la nature*. Bernard Jasmin, transl. Camille Limoges, introdruction and notes. Paris: J. Vrin, 171 pp.
Llewellyn, Alexander. 1972. *The Decade of Reform: The 1830s*. Newton Abbot: David & Charles, 221 pp.
Lousley, J.E., ed. 1951. *The Study of the Distribution of British Plants*. Arborath: Botanical Society of the British Isles.

———, ed. 1957. *Progress in the Study of the British Flora*. L: Botanical Society of the British Isles.

———. 1962. 'Some New Facts about the Early History of the Society', *Bot. Soc. Brit. Is. Proc.* 4:410–12.
Lyell, Charles. 1830–33. *Principles of Geology, Being an Attempt to Explain the Former Changes of the Earth's Surface by Reference to Causes Now in Operation*. L: J. Murray.
MacGillivray, William. 1832. 'Remarks on the Phenogamic [!] Vegetation of the River Dee, in Aberdeenshire', *Wernerian Nat. Hist. Soc. Mem.* 6:539–56.
MacLeod, Roy. 1965. 'Evolutionism and Richard Owen, 1830–68', *Isis* 56:259–80.

———. 1974. 'The Ayrton Incident: A Commentary on the Relations of Science and Government in England, 1870–1873', pp. 45–78 in Thackray & Mendelsohn 1974.

_____ & Collins, Peter, eds. 1981. *The Parliament of Science: The British Association for the Advancement of Science, 1831–1981*. Northwood, UF: Science Reviews.

Mabberley, D.J. 1985. *Jupiter Botanicus: Robert Brown of the British Museum*. Braunschweig: J. Cramer, 500 pp.

Manuel, Frank E. 1963. *Isaac Newton, Historian*. Cambridge, MA: Harvard University Press, viii + 328 pp.

_____. 1968. *A Portrait of Isaac Newton*. Cambridge, MA: Harvard University Press, xvii + 478 pp.

_____. 1974. *The Religion of Isaac Newton*. Oxford: Clarendon Press, vi + 141 pp.

Masterman, Margaret. 1970. 'The Nature of a Paradigm', pp. 59–89 in Lakatos & Musgrave 1970.

May, Caroline. 1990. *A Victorian Flora: The Flower Paintings of Caroline May*. Ed. Richard Mabey. Woodstock, NY: Overlook Press, 183 pp.

Mayr, Ernst. 1972a. 'Lamarck Revisited', *JHB* 5:55–94.

_____. 1972b. 'The Nature of the Darwinian Revolution', *Science* 176:981–9.

_____. 1982. *The Growth of Biological Thought: Diversity, Evolution, and Inheritance*. Cambridge, MA: Harvard University Press, xiv + 974 pp.

Mazlish, Bruce. 1975. *James and John Stuart Mill: Father and Son in the Nineteenth Century*. NYC: Basic Books, xii + 484 pp.

Mellersh, H.E.L. 1968. *FitzRoy of the 'Beagle'*. L: Rupert Hart-Davis, 307 pp.

Mercer, T.S. 1965. *Tales and Scandals of Old Thames Ditton*. Thames Ditton: T. S. Mercer, 56 pp.

_____. 1970. *More Thames Ditton Tales and Scandals*. Thames Ditton: Author, 80 pp.

_____. 1971. *A Souvenir of Old Thames Ditton*. Thames Ditton: T. S. Mercer, 56 pp.

Meyen, Franz J.F. 1846. *Outlines of the Geography of Plants: with Particular Enquiries concerning the Native Country, the Culture and the Uses of the Principle Cultivated Plants*. Margaret Johnston, transl. L: Ray Society, x + 422 pp. + chart.

Middleton, W.E. Knowles. 1969. *Invention of the Meteorological Instruments*. Baltimore: Johns Hopkins University Press, xiv + 362 pp.

Mill, John S. 1845. 'Note on the Species of Œnanthe', *P* 2:48–9, 116. Reprinted in Mill 1963–91, vol. 31, pp. 265–6.

_____. 1963–91. *The Collected Edition of the Works*, ed. John M. Robson et al., 33 vols. Toronto: University of Toronto Press. Includes: *The Earlier Letters, 1812–1848*, vols 12–13; *The Later Letters, 1849–1873*, vols 14–17; *Additional Letters*, vol. 32; 'Botanical Writings, 1840–61', vol. 31, pp. 257–320.

Millhauser, Milton. 1959. *Just before Darwin: Robert Chambers and Vestiges*. Middleton CT: Wesleyan U.P., ix + 248 pp.

Miller, David P. & Reill, Peter H., eds. 1996. *Visions of Empire: Voyages, Botany, and Representations of Nature*. Cambridge: Cambridge University Press, xix + 370 pp.

Mills, Eric L. 1984. 'A View of Edward Forbes, Naturalist', *Archives of Nat. Hist.* 11:365–93.

Mirbel, Charles F.B. de. 1825a. 'Essai sur la distribution géographique des Conifères', *Mus. Hist. nat. Mém.* 13:28–76.

_____. 1825b. 'Sur la distribution géographique de la famillie des Chénopodées', *Mus. Hist. nat. Mém.* 13:192–203.

_____. 1827. 'Recherches sur la distribution géographique des végétaux phanérogames dans l'Ancien Monde, depuis l'équateur jusqu'au pôle arctique', *Mus. Hist. nat. Mém.* 14:349–474.

Mitchell, Sally, ed. 1988. *Victorian Britain: An Encyclopedia.* NYC: Garland, xxi + 986 pp.

Moore, David & More, Alexander G. 1866. *Contributions towards a Cybele Hibernica, being Outlines of the Geographical Distribution of Plants in Ireland.* Dublin: Hodges & Smith, lvi + 399 pp.

_____. 1867. 'On the Climate, Flora, and Crops of Ireland', *Report of the Proceedings of the International Horticultural Exhibition and Botanical Congress Held in London, May 1866.* L: Truscott, Son & Simmons, pp.165–76.

Moore, James R. 1982. 'Darwin Lies in Westminster Abbey', *Linnean Soc. London Biol. J.* 17:97–113.

_____. 1985. 'Darwin's Genesis and Revelations', *Isis* 76:570–80.

_____, ed. 1988a. *Religion in Victorian Britain*, vol. 3: Sources. Manchester, University Manchester Press, xi + 545 pp.

_____. 1988b. 'Freethought, Secularism, Agnosticism: The Case of Charles Darwin', vol. 1, pp. 274–319 in Parsons 1988.

_____, ed. 1989. *History, Humanity and Evolution.* Cambridge: Cambridge University Press.

More, Alexander G. 1865. 'On the Distribution of Birds in Great Britain during the Nesting Season', *Ibis* 1:1–27, 119–42, 425– 58 + map; reprinted in More 1898, pp. 401–84 + map.

_____. 1898. *Life and Letters of Alexander Goodman More, F.R.S.E., F.L.S., M.R.I.A., with Selections from His Zoological and Botanical Writings.* Frances M. More, comp. and preface, C.B. Moffat, ed. Dublin: Hodges, Figgis, xii + 642 pp.

_____; Colgan, Nathaniel & Scully, Reginald W. 1898. *Contributions towards a Cybele Hibernica, being Outlines of the Geographical Distribution of Plants in Ireland.* Edn 2, Dublin: Edward Ponsonby, xxxvi + 538 pp.

Morley, John. 1887. 'Richard Cobden (1804–1865)', *DNB*.

Morrell, J.B. 1972. 'Science and Scottish University Reform: Edinburgh in 1826', *British J. Hist. Sci.* 6:39–56.

Motte, Jean. 1970. 'Pierre-Auguste-Marie Broussonet (1761–1807)', *DSB* 2:509–511.

Mullen, Shirley A. 1988. 'Freethought', pp. 312–14 in Mitchell 1988.
Munsell, F. Darrell. 1988. 'Irish Famine', p. 403 in Mitchell 1988.
Murray, Hugh et al. 1834. *An Encyclopaedia of Geography: Comprising a Complete Description of the Earth* ... L: Longman, 1567 pp.
Nelson, E. Charles. 1993a. 'Mapping Plant Distribution Patterns: Two Pioneering Examples from Ireland Published in the 1860s', *Arch. Natur. Hist.* 20:391–403.
———. 1993b. 'Searching the Archives for Botanists, With Some Irish Case Histories', *Huntia* 9:5–19.
[Newman, Edward]. 1844. 'Notice of *The London Journal of Botany.* By W.J. Hooker ... June 1844 ..'., *P* 1:1022–31.
[Newman, Thomas P.] 1876. *Memoir of the Life and Works of Edward Newman.* L: John Van Voorst, 32 pp. [Reprinted, E.W. Classey, introduction Faringdon: Classey, 1980.]
Nicolson, Malcolm. 1987. 'Alexander von Humboldt, Humboldtian Science, and the Origins of the Study of Vegetation', *Hist. Sci.* 25:167–94.
———. 1990. 'Alexander von Humboldt and the Geography of Vegetation', pp. 169–85 in Cunningham & Jardine 1990.
Nougarède, A. 1974. 'Charles François Brisseau de Mirbel (1776–1854)', *DSB* 9:418–19.
Novo, Laura. 1988. 'Elementary Education', pp. 241–3 in Mitchell 1988.
O'Byrne, William R. 1849. *A Naval Biographical Dictionary: Comprising the Life and Services of Every Living Officer in Her Majesty's Navy, from the Rank of Admiral of the Fleet to That of Lieutenant* ... L: J. Murray, ix + 1400 pp.
Oliver, Francis W., ed. 1913. *Makers of British Botany: A Collection of Biographies by Living Botanists.* Cambridge: Cambridge University Press, iv + 332 pp.
O'Rahilly, Ronan. 1948. *Benjamin Alcock, the First Professor of Anatomy and Physiology in Queen's College, Cork.* Oxford: Blackwell, 37 pp.
Orth, John V. 1988. 'Legal Profession', pp. 444–5 in Mitchell 1988.
Ospovat, Dov. 1981. *The Development of Darwin's Theory: Natural History, Natural Theology, and Natural Selection, 1838–1859.* NYC: Cambridge University Press, xii + 301 pp.
Page, L.E. 1972. 'John Fleming (1785–1857)', *DSB* 5:31–2.
Parshall, Karen H. 1982. 'Varieties as Incipient Species: Darwin's Numerical Analysis', *JHB* 15:191–214.
Parsons, Gerald, ed. 1988. *Religion in Victorian Britain.* Vol. 1: *Traditions.* Manchester: Manchester University Press.
Parssinen, T.M. Fall 1974. 'Popular Science and Society: The Phrenology Movement in Early-Victorian Britain', *J. Social Hist.* 8:1–20.
Payne, Joseph F. 1885. 'William Pulteney Alison (1790–1859)', *DNB*.
Peckham, Morse. 1959. *See* Darwin 1959c.
Pilet, Paul E. 1971a. 'Alphonse de Candolle (1806–93)', *DSB* 3: 42–43.

———. 1971b. 'Augustin-Pyramus de Candolle (1778–1841)', *DSB* 3:43–45.
Popper, Karl. 1962. *Conjectures and Refutations: The Growth of Scientific Knowledge*. NYC: Basic Books, xii + 412 pp.
———. 1965. *The Logic of Scientific Discovery*. Revised edn, NYC: Harper Torchbook, 480 pp.
Porter, Duncan M. 1980. 'Charles Darwin's Plant Collections from the Voyage of the *Beagle*', *J. Soc. Bibliog. Nat. Hist.* 9:515–25.
Porter, Roy. 1995. *London: A Social History*. Cambridge, MA: Harvard University Press, xvi + 431 pp.
Poynter, F.N.L., ed. 1958. *The History and Philosophy of Knowledge of the Brain and Its Functions*. Springfield, IL: Thomas.
Praeger, R. Lloyd. 1949. *Some Irish Naturalists: A Biographical Note-book*. Dundalk, Ireland: W. Tempest, Dundalgan Press, 208 pp.
Prideaux, Thomas S. 1840. *Strictures on the Conduct of H. Watson, F.L.S., in his capacity of Editor of the Phrenological Journal; With an Appendix, containing a Speculative Analysis of the Mental Functions*. Ryde UT: Hellyer, 72 pp.
Pritzel, Georg A. 1871–77. *Thesaurus Literaturae Botanicae*. Edn 2, Leipzig: Brockhaus, iii + 577 pp. Koenigstein: Koeltz, 1972.
Prothero, Rowland E. 1898a. 'Arthur Penryhn Stanley (1815–81)', *DNB*.
———. 1898b. 'Edward Stanley (1779–1849)', *DNB*.
———. 1898c. 'Owen Stanley (1811–50)', *DNB*.
Rae, William F. 1897. 'Lord John Russell (1792–1878)', *DNB*.
Ransford, Charles. 1846. *Biographical Sketch of the Late Robert Graham, M.D., F.R.S.E., Professor of Medicine and Botany in the University of Edinburgh*. Edinburgh: Harveian Society, 40 pp.
Rapson, Edward J. 1890. 'Charles Grant (1778–1866), Baron Glenelg', *DNB*.
[Reece, Richard]. 1803. Review of 'A Letter from Charles Villers to Georges Cuvier, Member of the National Institute of France, on a New Theory of the Brain, as the Intermediate Organ of the Intellectual and Moral Faculties, by Dr Gall', *Edinburgh Review* 2:147–60.
Rehbock, Philip F. 1979. 'Edward Forbes (1815–1854) – An Annotated List of Published and Unpublished Writings', *J. Soc. Bibliog. Nat. Hist.* 9:171–218.
———. 1983. *The Philosophical Naturalists: Themes in Early Nineteenth Century British Biology*. Madison: University Wisconsin Press, xv + 281 pp.
———. 1985. 'John Fleming (1785–1857) and the Economy of Nature', pp. 129–40 in Wheeler & Price 1985.
Reilly, Desmond. 1955. 'Robert John Kane (1809–90): Irish Chemist and Educator', *J. Chem. Education* 32:404–6.
Reingold, Nathan. 1980. 'Through Paradigm-Land to a Normal History of Science', *Social Studies of Science* 10:475–96.

Reynolds, John. 1847. 'Testimonial to Mr. G.E. Dennes', *P* 2:815.
Richards, Graham. 1992. *Mental Machinery: The Origins and Consequences of Psychological Ideas. Part 1: 1600–1850*. Baltimore MD: Johns Hopkins University Press, xi + 499 pp.
Richards, Robert J. 1987. *Darwin and the Emergence of Evolutionary Theories of Mind and Behavior*. Chicago: University of Chicago Press, xvii + 700 pp.
Richardson, R. Alan. 1981. 'Biogeography and the Genesis of Darwin's Ideas on Transmutation', *JHB* 14:1–41.
Rigg, James M. 1899. 'Richard Whately (1787–1863)', *DNB*.
Roger, Alain & Guéry, François, eds. 1991. *Maitres et protecteurs de la nature*. Seyssel: Champ Vallon.
Rosner, Lisa. 1991. *Medical Education in the Age of Improvement: Edinburgh Students and Apprentices, 1760–1826*. Edinburgh: Edinburgh University Press, viii + 273 pp.
Royal Society of London. 1867–1925. *Catalogue of Scientific Papers, 1800–1900*. 19 vols. in four series. L and Cambridge: C.J. Clay & Sons, Cambridge University Press.
Royle, Edward. 1971. *Radical Politics, 1790–1900: Religion and Unbelief*. L: Harlow, Longman, viii + 152 pp.
_____. 1974. *Victorian Infidels: The Origins of the British Secularist Movement, 1791–1866*. Manchester: University of Manchester Press, viii + 357 pp.
Rudwick, Martin J. 1974. 'Darwin and Glen Roy: A "Great Failure" in Scientific Method', *Stud. Hist. Philos. Sci.* 5:97–185.
Rumball, J.R. 1838. 'Letter on the Organ of Wit', *PJ* 11:391–3.
Runyan, William M. 1982. *Life Histories and Psychobiography: Explorations in Theory and Method*. NYC: Oxford University Press, xiii + 288 pp.
Rupke, Nicolaas A. 1994. *Richard Owen: Victorian Naturalist*. New Haven: Yale University Press, xviii + 462 pp.
Ruse, Michael. 1975. 'The Relationship between Science and Religion in Britain, 1830–70', *Church Hist.* 44:505–522.
_____. 1979. *The Darwinian Revolution: Science Red in Tooth and Claw*. Chicago: University of Chicago Press, xvi + 320 pp.
Russell-Gebbett, Jean. 1977. *Henslow of Hitcham: Botanist, Educationalist and Clergyman*. Lavenham: Terence Dalton, 139 pp.
Sayre, Anne. 1975. *Rosalind Franklin and DNA*. NYC: Norton, 221 pp.
Schwartz, Joel S. 1974. 'Charles Darwin's Debt to Malthus and Edward Blyth', *JHB* 7:301–18.
_____. 1980. Review of Eiseley 1979, *Isis* 71:517.
_____. 1990. 'Darwin, Wallace, and Huxley, and *Vestiges of the Natural History of Creation*', *JHB* 23:127–53.

Schweber, Silvan S. 1989. 'John Herschel and Charles Darwin: A Study in Parallel Lives', *JHB* 22:1–71.

Scott, William. 1827a. 'Of Wit and the Feeling of the Ludicrous', *PJ* 4:195–242.

――― [anon.]. 1827b. 'On the Faculty of Comparison', *PJ* 4:319–63.

―――. 1836. *The Harmony of Phrenology with Scripture: Shewn in a Refutation of the Philosophical Errors contained in Mr Combe's 'Constitution of Man'*. Edinburgh: Fraser, 332 pp.

Secord, Anne. 1994. 'Corresponding Interests: Artisans and Gentlemen in Nineteenth-Century Natural History', *Brit.J.Hist. Sci.* 27:383–408.

Secord, James A. 1991. 'Edinburgh Lamarckians: Robert Jameson and Robert E. Grant', *JHB* 24:1–18.

Seubert, Moritz. 1844. *Flora Azorica quam ex collectionibus schedisque Hochstetteri patris et filii* ... Bonnae: A. Marcum, vi + 50 pp. + 15 plts.

――― & Hochstetter, Christian F. 1843. 'Übersicht der Flora der azorischen Inseln', *Archiv für Naturgeschichte* 9:1–24 + sketch.

Shapin, Steven. 1975. 'Phrenological Knowledge and the Social Structure of Early 19th-Century Edinburgh', *AS* 32:219–43.

Sheets-Pyenson, Susan. 1981a. 'A Measure of Success: The Publication of Natural History Journals in Early Victorian Britain', *Publishing Hist.* 9:21–36.

―――. 1981b. 'Darwin's Data: His Reading of Natural History Journals, 1837–1842', *JHB* 14:231–48.

Shortland, Michael & Yeo, Richard, eds. 1996. *Telling Lives in Science: Essays on Scientific Biography*. Cambridge: Cambridge University Press, xi + 295 pp.

Shteir, Ann B. 1987. 'Botany in the Breakfast Room: Women and Early Nineteenth-Century British Plant Study', chap. 2 in Abir-am & Outram 1987.

Sidebotham, Joseph. 1842. 'Note on the Oxlips from Bardfield', *P* 1:238.

――― 1844. 'Remarks on the London List of British Plants', *P* 1:972– 4.

―――. 1846. 'Supposed Transformation of Oats into Rye', *P* 2:589– 90.

―――. 1847. 'Note on Raising Primulas from Seed', *P* 2:887.

―――. 1848a. 'Remarks on Certain "Excluded Species" Placed at the End of *The London Catalogue*', *P* 3:70–1.

―――. 1848b. 'Further Remarks on the Second Edition of *The London Catalogue of British Plants*', *P* 3:140–1.

―――. 1848c. 'Reply to Mr. Watson's Observations, *Phytol.* iii. 84', *P* 3:188. [Newman's editorial comment, p. 189.]

―――. 1849. 'Experiments on the Specific Identity of the Cowslip and Primrose', *P* 3:703–5.

Smit, Pieter. 1976a. 'Gottfried Reinhold Treviranus (1776–1837)', *DSB* 13:460–2.

―――. 1976b. 'Ludolph Christian Treviranus (1779–1864)', *DSB* 13:462–3.

Smith, Fabienne. 1990. 'Charles Darwin's Ill Health', *JHB* 23:443– 59.

Sonstegard, David A., Matthews, Larry S. & Kaufer, Herbert. January 1978. 'The Surgical Replacement of the Human Knee Joint', *Sci. Amer.* 238, no. 1:44–51.

Spall, Richard F., Jr. 1988. 'Anti-Corn-Law League (1838–1846)', p. 29 in Mitchell 1988.

Spencer, Herbert. 1904. *An Autobiography.* 2 vols. NYC: Appeleton.

Spurzheim, Johann C. 1815. *The Physiognomical System of Drs Gall and Spurzheim.* Baldwin, Cradock & Joy, xviii + 581 pp.

———. 1817. *Examination of the Objections Made in Britain against the Doctrines of Gall and Spurzheim.* Edinburgh

Stafleu, Frans A. & Cowan, Richard S. 1976–92. *Taxonomic Literature: A Selective Guide to Botanical Publications and Collections with Dates, Commentaries and Types.* Edn 2, 8 vols. Utrecht: The Netherlands Bohn, Scheltema & Holkema.

Stanley, Edward. 1836?. *Heads for the Arrangement of Local Information in Various Departments of Parochial and Rural Interest.* London: Hatchard & Son, 31 pp. Edn 2, 1848.

Stearn, William T. 1970. 'Robert Brown (1773–1858)', *DSB* 2:516– 22.

———. 1973. 'John Lindley (1799–1865)', *DSB* 8:371–3.

———. 1981. *The Natural History Museum at South Kensington: A History of the British Museum (Natural History), 1753–1980.* L: Heinemann, xxiii + 414 pp.

Stephen, Leslie. 1890. 'Sir William Hamilton (1788–1856)', *DNB*.

Stephens, Michael D. & Roderick, Gordon W. 1972. 'Nineteenth Century Ventures in Liverpool's Scientific Education', *AS* 28:61–86.

Stephens, W.B., ed. 1970. *History of Congleton.* Manchester: Manchester University Press, xviii + 365 pp.

Stone, Irving. 1980. *The Origin: A Biographical Novel of Charles Darwin.* Garden City NY: Doubleday, 743 pp.

Sulloway, Frank J. 1982a. 'Darwin and His Finches: The Evolution of a Legend', *JHB* 15:1–53.

———. 1982b. 'Darwin's Conversion: The *Beagle* Voyage and Its Aftermath', *JHB* 15:325–96.

Swazey, Judith P. 1970. 'Action Propre and Action Commune: The Localization of Cerebral Function', *JHB* 3:213–34.

Sweet, Jessie M. 1970a. 'William Bullock's Collection and the University of Edinburgh, 1819', *AS* 26:23–32 + 3 plts.

———. 1970b. 'The Collection of Louis Dufresne (1752–1832)', *AS* 26:33–71 + 3 plts.

Tansley, Arthur G., ed. & part author. 1911. *Types of British Vegetation.* Cambridge: Cambridge University Press, xx + 416 pp.

———. 1939. *The British Islands and Their Vegetation.* Cambridge: Cambridge University Press, xxxviii + 930 pp.

Taylor, George. 1970. 'George Bentham (1800–84)', *DSB* 1:614–15.

Temkin, Owsei. 1947. 'Gall and the Phrenological Movement', *Bull. Hist. Med.* 21: 275–321.

———. 1953. 'Remarks on the Neurology of Gall and Spurzheim', vol. 2, pp. 282–9 in Underwood 1953.

Tenon, Jacques R., Portal, A. Sabatier, R., Pinel, P., and Cuvier, G. 1809. 'Report on a Memoir of Drs Gall and Spurzheim, relative To the Anatomy of the Brain', *Edinb. Med. & Surg. J.* 5:36–66.

Thackray, Arnold. 1974. 'Natural Knowledge in Cultural Context: The Manchester Model', *Amer. Hist. Rev.* 79:672–709.

——— & Mendelsohn, Everett, eds. 1974. *Science and Values: Patterns of Traditon and Change*. NYC: Humanities P.

Thurmann, Jules. 1850. 'Summary of *Essai de phytostatistique*', transl. Hewett C. Watson, *P* 3:918–21.

Trevelyan, Raleigh. 1978. *A Pre-Raphaelite Circle*. L: Chatto & Windus, 256 pp.

Treviranus, Gottfried R. 1802–21. *Biologie, oder Philosophie der lebenden Natur für Naturforscher und Aerzte*. 6 vols. Göttingen: J.F. Röwer.

[Trimen, Henry]. 1868–70. Review of Watson 1868–70. *J. Bot.* 6:374–7; 7: 368–9; 8:394–7.

Tuckey, J.H. 1818. *Narrative of an Expedition to Explore the River Zaire*. L: Murray.

Turner, Dawson & Dillwyn, Lewis W. 1805. *The Botanist's Guide through England and Wales*. 2 vols. L: Phillips & Fardon.

Turner, Frank M. 1993. *Contesting Cultural Authority: Essays in Victorian Intellectual Life*. NYC: Cambridge University Press, xiv + 368 pp.

Turrill, William B. 1959. *The Royal Botanic Gardens Kew: Past and Present*. L: H. Jenkins, 256 pp.

———. 1963. *Joseph Dalton Hooker: Botanist, Explorer, and Administrator*. L: Thomas Nelson, xi + 228 pp.

Tutin, T. 1953. 'The Vegetation of the Azores', *J. Ecol.* 41:53–61.

Underwood, Edgar A. 1951. 'The History of the Quantitative Approach in Medicine', *Bull. Hist. Med.* 7:265–74.

———, ed. 1953. *Science, Medicine and History*. 2 vols, NYC: Oxford University Press.

Valentine, David H. 1947. 'Studies in British Primulas. I: Hybridization between Primrose and Oxlip (*Primula vulgaris* Huds. and *P. elatior* Schreb.)', *New P* 46:229–53.

———. 1948. 'Studies in British Primulas. II: Ecology and Taxonomy of Primrose and Oxlip (*Primula vulgaris* Huds. and *P. elatior* Schreb.)', *New P* 47:111–30.

———. 1952. 'Studies in British Primulas. III: Hybridization between *Primula elatior* (L.) Hill and *P. veris* L'. *New P* 50:383–99 + 2 plts.

———. 1953. 'Evolutionary Aspects of Species Differences in *Primula*', *Soc. Exper. Biol. Symposium* 7:146.

_____. 'Studies in British Primulas. IV: Hybridization between *Primula vulgaris* Huds. and *P. veris* L'. *New P* 54:70–80.

_____. 1961. 'Evolution in the Genus *Primula*', pp. 71–87 in Wanstall 1961.

_____. 1966. 'The Experimental Taxonomy of Some *Primula* Species', *Bot. Soc. Edinburgh Trans.* 40:169–180.

von Hofsten, Nils. 1916. 'Zur Geschichte des Diskontinuitätsproblems in der Biogeographie', *Zoologische Annalen* 7:197–353.

Vorzimmer, Peter J. 1970. *Charles Darwin, The Years of Controversy: The Origin of Species and Its Critics, 1859–1882*. Philadelphia PA: Temple University Press, xx + 300 pp.

_____. 1977. 'The Darwin Reading Notebooks (1838–1860)', *JHB* 10: 107–53.

Wach, Howard M. 1988. 'Culture and the Middle Classes: Popular Knowledge in Industrial Manchester', *J. British Studies* 27: 375–404.

Wagenitz, Gerhard. 1972. 'August Heinrich Rudolf Grisebach', *DSB* 5:546–7.

Wahlenberg, Göran. 1812. *Flora Lapponica exhibens Plantas Geographice ...* Berlin: Taberna Libraria Scholae Realis, lxvi + 550 pp.

_____. 1813. *De Vegetatione et Climate in Helvetia septentrionali ...* Turici Helvetorum: Orell, Fuessli, xcviii + 200 pp.

_____. 1814. *Flora Carpatorum ... cum Mappa physico-geographica, Tabula altitudinem ...* Göttingen: Vandenhöck & Ruprecht, cxviii + 409 pp.

Walsh, Anthony A. 1970. 'Is Phrenology Foolish? A Rejoinder', *J. Hist. Behav. Sci.* 6:358–61.

_____. 1971. 'George Combe: A Portrait of a heretofore generally Unknown Behaviorist', *J. Hist. Behav. Sci.* 7:269–78.

_____. 1975. 'Johann Christoph Spurzheim (1776–1832)', *DSB* 12:596–7.

Walters, Stuart M. 1951. 'The Study of Plant Distribution', pp. 12–23 in Lousley 1951.

_____. 1957. 'Distribution Maps of Plants – An Historical Survey', pp. 89–96 in Lousley 1957.

_____. 1981. *The Shaping of Cambridge Botany: A Short History of Whole-Plant Botany in Cambridge from the Time of Ray into the Present Century*. Cambridge: Cambridge University Press, xv + 121 pp.

Wanstall, P.J., ed. 1961. *A Darwin Centenary*. L: Botanical Society of the British Isles.

_____, ed. 1963. *Local Floras*. L: Botanical Society of the British Isles.

Watson, James D. 1980. *The Double Helix: A Personal Account of the Discovery of the Structure of DNA. A Norton Critical Edition: Text, Commentary, Reviews, Original Papers*, ed. Gunther S. Stent. NYC: Norton, xxvii + 298 pp.

Webb, Philip B. & Berthelot, Sabin. 1836–50. *Histoire naturelle des îsles canariensis*. 3 vols in 9 + atlas. Paris: Béthune-Mellier.

Weissenborn, W. 1838a. 'On Spontaneous Generation', *MNH* 2:369–81, 621–4.

_____. 1838b. 'On the Transformation of Oats into Rye', *MNH* 2:670–672.

Wells, George A. 1972. 'Johann Wolfgang von Goethe (1749–1832)', *DSB* 5:442–6.
Westfall, Richard S. 1980. *Never at Rest: A Biography of Isaac Newton*. NYC: Cambridge University Press, xviii + 908pp.
Wheeler, Alwyne & Price, James H., eds. 1985. From *Linnaeus to Darwin: Commentaries on the History of Biology and Geology*. L: Society for the History of Natural History.
Wheeler, Thomas S. 1945. 'Sir Robert Kane', *Endeavour* 4:91–3.
Williams, Carrington B. 1943. 'Area and Number of Species', *Nature* 152:264–7.
———. 1964. *Patterns in the Balance of Nature and Related Problems In Quantitative Ecology*. NYC: Academic Press, 324 pp.
Williams, Weslely C. 1972. 'Thomas Henry Huxley (1825–95)', *DSB* 6:589–97.
———. 1974. 'Richard Owen (1804–92)', *DSB* 10:260–3.
Willis, John C. et al. 1922. *Age and Area: A Study in Geographical Distribution and Origin of Species*. Cambridge: Cambridge University Press, x + 259 pp.
Wilson, George & Geikie, Archibald. 1861. *Memoir of Edward Forbes, F.R.S., late of Edinburgh, Regius Professor of Natural History* ... Cambridge: Macmillan, 589 pp.
Wilson, Leonard G. 1973. 'Charles Lyell (1797–1875)', *DSB* 8:563–76.
Wilson, William. 1846. 'Thoughts on the Progressive Development of Species', *P* 2:444–7.
Winch, Nathaniel J. 1819. *An Essay on the Geographical Distribution of Plants, through the Counties of Northumberland, Cumberland, and Durham*. Newcastle upon Tyne: E. Charnley, 52 pp. Edn 2, 1825, 54 pp.
Winslow, John H. 1971. *Darwin's Victorian Malady: Evidence for Its Medically Induced Origin*. Philadelphia, PA: American Philosophical Society, vi + 94 pp. [*Memoirs*, vol. 88.]
Woodbridge, George. 1970. *The Reform Bill of 1832*. NYC: Crowell, vii + 104 pp.
Woodham-Smith, Cecil. 1962. *The Great Hunger: Ireland, 1845–1849*. NYC: Harper & Row, 510 pp.
Woodward, Ernest L. 1962. *The Age of Reform, 1815–1870*. Edn 2, NYC: Oxford University Press, xx + 644 pp.
Yeo, P.F. 1973. 'The Azorean Species of Euphrasia', *Boletim do Museu Municipal do Funchal*, no. 27:74–83.
Young, Robert M. 1970. *Mind, Brain and Adaptation in the Nineteenth Century: Cerebral Localization and Its Biological Context from Gall to Ferrier*. NYC: Oxford University Press, 136 pp.
———. 1972. 'Franz Joseph Gall (1758–1828)', *DSB* 5:250–6.
Zastoupil, Lynn. 1988. 'Indian Mutiny (1857)', pp. 390–1 in Mitchell 1988.

Index

Alison, William P. 15, 23
Allman, George J. 114
Arnott, George A.W. 110, 114–15, 221, 223, 251
Ayrton, Acton S. 213

Babington, Charles C. 105, 110, 112, 113, 121, 127, 136, 139, 153, 212, 225, 235, 250, 252
  Manual of British Botany (1843) 129, 132–3, 140, 155
  relations with Watson 128–33
Bagnall, James E. 227, 252
Baker, John G. 42, 47, 214–18, 221, 222, 226, 227, 228, 229, 230, 236, 239, 251
  portrait 217
Balfour, J. Hutton 104, 111, 115, 121, 127, 134–6, 139, 225, 235, 252
  relations with Watson 29, 128–33
Banks, Joseph 99
Barrett, William B. 47
Bell, Isaac 11
Bentham, George 208, 212, 222, 251
Bonpland, Aimé 24, 27
Britten, James 218, 251
Brown, Robert 27–8, 32, 104

Camerini, Jane 3
Candolle, Alphonse L.P.P. 181, 184–5, 187–89, 216, 219, 221, 223–4, 236, 252
Candolle, Augustin P. de 26, 101, 234, 237
Cameron, Douglas G. 14, 15, 20
Chambers, Robert 16, 75, 148, 150, 153–4, 159
Cobden, Richard 73–4, 77
Combe, Andrew 50, 58–9, 235
  illness 46
  medical education 19
  phrenology writings 19
  teaches phrenology 23
Combe, George 50–51, 55–6, 59, 70, 73, 75–80, 117–18, 150, 160, 234, 235, 237, 252

early life 17
engagement 44
general writings 47, 68–70, 77, 79–80, 159
marriage 46
phrenology writings 19–20, 55, 67–8, 149, 158
portraits 18, 78
recommending Watson 29
studies phrenology 19
teaches phrenology 23
Comber, Thomas R. 227
Cox, Marion 43, 44, 46, 48, 235
Cox, Robert 58, 65

Dalton, John 22
Darwin, Charles R. 1, 88, 126, 147, 148, 161–175, 177–83, 190–93, 197–201, 203, 207, 211, 224, 230–31, 236, 237, 239, 250
  Azores visit 167, 178
  Edinburgh University, at 29–30, 33, 164
  On the Origin of Species 179, 191, 200, 203, 208
  portrait 170
  visits Watson 181
Dawson, E. 41, 81, 251
De la Beche, Henry 104–5
Dennes, George E. 138–40, 143, 145
Dickie, George 104, 119–20
Dilwyn, Lewis W. 27
Don, David 111
Drummond, Thomas 32
Duncan, William H. 112

Eastmond, Grace 46–7, 229–30, 235
Eveleigh, Joseph 12

FitzRoy, Robert 16, 88, 165
Fleming, John 220
Flower, Thomas B. 47
Forbes, Edward 33, 35, 98, 105, 113, 121–8, 158, 171–2, 174, 183, 211, 224, 233

281

Gall, Franz J. 16, 17, 21, 25–6
Godman, Frederick D. 204–6
Graham, Robert 30–31, 104, 111, 122, 129, 234
Gray, John E. 112, 134, 138, 143, 145, 221, 251
Gregory, William 68
Greville, Robert K. 111
Grisebach, August H.R. 184
Guthnick 8, 97

Hamilton, William 19
Hancock, George 55
Harvey, William H. 114
Henfrey, Arthur 184
Henslow, John 57, 112, 128–9, 165, 168, 223, 251
Herschel, John 68, 234, 237
Hochstetter, Christian F. 95, 205
Hodgson, William B. 59–60
Home, James 30
Hooker, Joseph 99–100, 103–7, 125, 168, 169, 171–3, 178–81, 184, 192, 208–14, 216, 222, 224, 230, 239
    portraits 106, 210, 250, 251
    The Student's Flora of the British Isles (1869) 212–13
Hooker, William J. 99–110, 111, 112, 113, 137, 223, 239, 251
    death 211
    Flora Scotica (1821) 29
    portraits 102, 108
    Professor at Glasgow 29, 31, 34
    The British Flora (1830 and later) 29, 31, 101, 110, 129, 132–3, 140, 155, 235
    visits Watson 43
    Watson's patron 37, 86, 116
Humboldt, F.W.H. Alexander von 1, 24–7, 32, 36, 37, 165, 234
Hunt, T. Carew 97, 180, 205

Irvine, Alexander 144, 221

Jackson, B. Daydon 227, 230
Jameson, Robert 28, 30, 147–8

Lecoq, Henri 184
Lemann, Charles M. 97
Lewes, George H. 17, 21, 23

Lindley, John 31, 32, 140, 148, 155, 190, 192, 220
Linnaeus, Carl 24, 26
Lloyd, John 8
Lloyd, John H. 50, 116
Lloyd, Louisa M. (Mrs Napier) 50–53
Luxford, George 139

McClelland, James 61
MacGillivray, William 28–9, 34
Mackenzie, George 71
Meyen, Franz J.F. 184
Mill, John S. 134, 137, 142, 201, 203
Miller, H. David G. 94
Moore, David 219
More, Alexander G. 218–20, 221

Napier, William 51–3
Newbould, W.W. 218, 227, 230
Newman, Edward 139, 140, 142–3, 214, 235

Parfett, Alice Ada 47
Peel, Robert 8, 76, 251
Prideaux, Thomas 62–3, 233

Richardson, John 37
Roscoe, William 13–14
Royle, John F. 113
Rumball, J. R. 55
Russel, Alexander 75
Russell, Frances 76
Russell, John 76

Scholes, Thomas S. 22, 112
Scott, William 20, 23, 56–7, 149
Seemann, Carl 221–2
Seubert, Moritz 95, 98
Shaen, Richard 75
Sidebotham, Joseph 142, 154, 200
Simpson, James 72
Smith, Christen 28
Spurzheim, Johann C. 16, 17, 21
Stanley, Edward 11, 12, 57, 115, 118

Tansley, Arthur G. 35
Theophrastos 24
Thurmann, Jules 109
Trevelyan, Walter C. 49, 61, 115, 252
Treviranus, Gottfried R. 32

Trimen, Henry 222–3, 233
Turner, Dawson 27, 99, 103

Vidal, Alexander T.E. 86, 88, 89, 93, 98

Wahlenberg, Göran 26, 34
Wakefield, Gilbert 41, 116
Watson, Caroline (Mrs Lloyd) 50
Watson, Harriett Anne (Mrs Scholes) 22
Watson, Harriett Powell (mother) 7–9
Watson, Hewett C.
  Azores
    plant studies 89–93, 204–8
    trip 86–95
  birth 7
  Cybele Britannica (1847–59 and later supplements) 145, 179, 183, 188–90, 192, 199, 203, 213, 220, 222
  editor of *Phrenological Journal* 58–60, 62–5
  exchange club idea 31
  Geographical Distribution of British Plants (1843) 98
  Gold Medal Prize for Physiological Botany (1831) 32–3
  legal apprenticeship 12
  New Botanist's Guide to the Localities of the Rarer Plants (1835–7) 86, 100, 122
  'On the Heads of Botanists' (1833) 56
  Outlines of Geographical Distribution of British Plants (1832) 26, 34–6, 130, 147
  phrenology
    at BAAS 61–2
    begins studies 20–23
    lectures on 60
  plant geography, begins studies 32–7
  portraits ix, 45, 143, 145, 202, 229
  Public Opinion; or Safe Revolution (1848) 74–5
  Remarks on the Geographical Distribution of British Plants (1835) 31, 84–5, 123, 185
  schools 11
  Statistics of Phrenology (1836) 22, 57, 67
  Thames Ditton home 42
  Topographical Botany (1873–4 & edn 2, 1883) 218, 227
  University of Edinburgh, at 15
Watson, Holland (father) 7–9, 11–13, 22, 42, 234, 251
Watson, Holland (son) 8
Watson, Julia (Mrs Gorton) 41
Watson, Louisa J. (Mrs Wakefield) 41
Watson, Zeph 8, 62
Webb, Philip B. 97
Wilson, George 31
Winch, Nathaniel J. 28, 34, 42, 84–5, 148, 251

For Product Safety Concerns and Information please contact our EU representative GPSR@taylorandfrancis.com
Taylor & Francis Verlag GmbH, Kaufingerstraße 24, 80331 München, Germany

www.ingramcontent.com/pod-product-compliance
Lightning Source LLC
Chambersburg PA
CBHW050553170426

43201CB00011B/1673